Mössbauer Effect
Methodology

Volume 9

MÖSSBAUER EFFECT METHODOLOGY

Proceedings of Symposia Sponsored by
the New England Nuclear Corporation, Boston

Edited by Irwin J. Gruverman

Volume 1
Proceedings of the First Symposium–January 1965

Volume 2
Proceedings of the Second Symposium–January 1966

Volume 3
Proceedings of the Third Symposium–January 1967

Volume 4
Proceedings of the Fourth Symposium–January 1968

Volume 5
Proceedings of the Fifth Symposium–February 1969

Volume 6
Proceedings of the Sixth Symposium–January 1970

Volume 7
Proceedings of the Seventh Symposium–January 1971

Volume 8
Proceedings of the Eighth Symposium–January 1973

Volume 9
Proceedings of the Ninth Symposium–February 1974

A Continuation Order Plan is available for this series. A continuation order will bring delivery of each new volume immediately upon publication. Volumes are billed only upon actual shipment. For further information please contact the publisher.

A Publication of the New England Nuclear Corporation

Mössbauer Effect Methodology

Volume 9

Edited by

Irwin J. Gruverman
Nuclear Medicine and Technology Center
New England Nuclear Corporation
Billerica, Massachusetts

and

Carl W. Seidel and David K. Dieterly
Nuclides and Sources Division
New England Nuclear Corporation
Billerica, Massachusetts

PLENUM PRESS • NEW YORK AND LONDON

The Library of Congress cataloged the first volume of this title as follows:

Symposium on Mössbauer Effect Methodology.
Mössbauer effect methodology; proceedings. 1st–
;1965–
ⱼBoston?ⱼ New England Nuclear Corporation; distributed
by Plenum Press, New York.

v. illus. 24 cm.

Symposia for 1965–   sponsored by the New England
Nuclear Corporation and the Technical Measurement Corporation.

1. Mössbauer effect—Addresses, essays, lectures.   ɪ. New England Nuclear Corporation, Boston.   ɪɪ. Technical Measurement Corporation.   ɪɪɪ. Title.

QC490.S94   65—21188

Library of Congress   ⱼ66ⱼ4ⱼ

Proceedings of the Ninth Symposium on Mössbauer
Effect Methodology held in Chicago, Illinois,
February 3, 1974

Library of Congress Catalog Card Number 65-21188

ISBN 978-1-4684-0939-0   ISBN 978-1-4684-0937-6 (eBook)
DOI 10.1007/978-1-4684-0937-6

© 1974 New England Nuclear Corporation
Softcover reprint of the hardcover 1st edition 1974

Plenum Press, New York
A Division of Plenum Publishing Corporation
227 West 17th Street, New York, N.Y. 10011

United Kingdom edition published by Plenum Press, London
A Division of Plenum Publishing Company, Ltd.
4a Lower John Street, London, W1R 3PD, England

Preface

This is the ninth volume of a continuing series intended to provide a forum for publication of developments in Mössbauer effect methodology and in spectroscopy and its applications.

Mössbauer Effect Methodology, Volume 9, records the proceedings of the Ninth Symposium on Mössbauer Effect Methodology. The symposium was sponsored by the New England Nuclear Corporation and interest was concentrated on spectroscopy and applications, with more than usual emphasis on new methodology. The symposium was held in the Palmer House in Chicago on February 3, 1974. Dr. Stanley Hanna presided over the afternoon and evening sessions.

Attendance was lower than usual; about one hundred participants were present. This may reflect the continuing pressure of travel budget limitations. Contributing Sponsors were Austin Science Associates, El Scint, Inc., Nuclear Science Instruments and Ranger Electronics. These organizations demonstrated their products for Mössbauer applications. The continuing improvements in the spectrometers and their adjuncts was evident.

The Selection Committee again had a most difficult task, and was obliged to accept only about half of the submitted papers. A most interesting group of papers on applications and spectroscopy featured reports on electronic relaxation phenomena, magnetic phase and spin transformations, photochromism in strontium titanate, lattice studies, and phase determination by Kossel analysis. The excellent methodology session included presentations on data analysis techniques for spectral folding, hyperfine interaction analysis and recoil-free fraction measurement, a backscatter spectrometer and a report on a Selective Excitation Double Mössbauer method to study time-dependent phenomena.

The editors are indebted to their colleagues, as always, for the fine effort in arriving at the final program after reviewing and discussing at length the large number of submitted papers. Dr. Hanna continued the tradition of coping with the extended program in excellent fashion. Robert MacKay and Nancy Snook extended organization and secretarial assistance, respectively.

The series is expected to continue.

I.G.
C.S.
D.D.

Billerica, Massachusetts
April, 1974

Contents

APPLICATIONS AND SPECTROSCOPY

APPLICATIONS
AND
SPECTROSCOPY

57Fe MÖSSBAUER EFFECT STUDIES OF HIGH-SPIN($^5T_2$) ⇌ LOW-SPIN($^1A_1$) TRANSITIONS IN ORGANIC COMPLEXES OF IRON(II)*

E. König† and G. Ritter

Institutes of Physical Chemistry II and
Physics II, University of Erlangen-Nürnberg,
D-8520 Erlangen, Germany

I. INTRODUCTION

The octahedral $3d^6$ configuration of transition metal ions is remarkable in that the so-called crossover situation may arise, provided the crystal-field energy, 10 Dq, and the Coulomb repulsion energy, approximately $\frac{5}{2}B + 4C$, are comparable in magnitude. If quasi-isolated molecules in solution are considered, an equilibrium between the high-spin($^5T_2$) and low-spin($^1A_1$) ground states may be rapidly established. At present, detailed information is available only for a single system, values of the rate constants for the two processes indicated by $^5T_2 \rightleftarrows {}^1A_1$ having been determined to be ~10^7 sec^{-1} at 298°K.[1] In solid systems, the behavior is modified by interactions within the lattice, thermally driven transitions between, and the coexistence of, high-spin ($^5T_2$) and low-spin($^1A_1$) states then being observed. If the effective symmetry is lower than cubic, the essential features of the $^5T_2 \rightleftarrows {}^1A_1$ transition are not significantly altered and, therefore, we will retain below the cubic designation of the states. Recently, thermally induced $^5T_2 \rightleftarrows {}^1A_1$ transitions have been studied in various systems of the [Fe^{II}-N_6] type. The six N atoms surrounding the central iron(II) ion are supplied, in

*Supported by the Deutsche Forschungsgemeinschaft and the Bundesministerium für Bildung und Wissenschaft.
†To whom correspondence should be addressed.

3

this case, by suitable organic ligands which additionally provide a high magnetic dilution of the solid. Thus the $^5T_2 \rightleftharpoons {}^1A_1$ transition may be conveniently isolated from other relevant interactions in the solid.

Various physical methods have been applied to the study of $^5T_2 \rightleftharpoons {}^1A_1$ transitions in complexes of iron(II) and among these, studies of the Mössbauer effect are of fundamental significance. Unlike electron paramagnetic resonance, e.g., its use is not restricted by the diamagnetism of the $^1A_1$ ground state or by the large zero-field splitting of the $^5T_2$ ground state in conjunction with a short spin-lattice relaxation time. Most importantly, the Mössbauer effect provides information about the $^5T_2$ and $^1A_1$ ground states separately, thus allowing detailed and quantitative studies of $^5T_2 \rightleftharpoons {}^1A_1$ transitions.

In general, two characteristic forms of behavior may be distinguished and these will be illustrated by one example for each. The compounds to be considered are $Fe(4,7-(CH_3)_2-phen)_2(NCS)_2$ and $Fe(papt)_2$. Here, $4,7-(CH_3)_2$-phen denotes the bidentate ligand 4,7-dimethyl-1,10-phenanthroline and papt the tridentate ligand 2-(2-pyridylamino)-4-(2-pyridyl)thiazolate.

4,7-(CH$_3$)$_2$-phen   papt

In the following, each of the physical properties of interest will be discussed for the two compounds together

II. MAGNETIC PROPERTIES

In the solid complex $Fe(4,7-(CH_3)_2-phen)_2(NCS)_2$, an abrupt change of the relevant magnetic properties is observed at the transition temperature, $T_c = 121.5°K$. This is illustrated in Fig. 1a by the reciprocal molar magnetic susceptibility $1/\chi_m^{corr}$ (corrected for the diamagnetism of the constituents) and by the effective

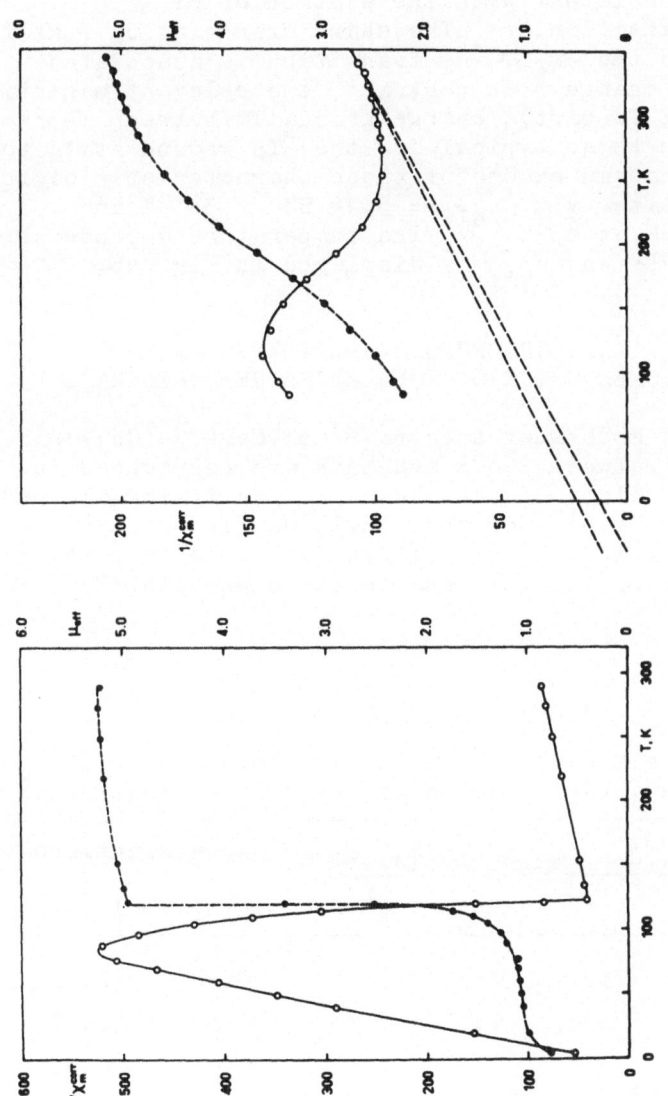

Fig. 1a and 1b. Temperature dependence of $1/\chi_m^{corr}$ (open points, left ordinate, in mol/emu) and of μ_{eff} (full points, right ordinate, in BM) for Fe(4,7-$(CH_3)_2$-phen)$_2$(NCS)$_2$ (left, a) and Fe(papt)$_2$ (right, b).

magnetic moment μ_{eff} where $\mu_{eff} = 2.828\sqrt{\chi_m^{corr} \cdot T}$. The values at the highest and lowest temperature studied, viz. $\mu_{eff} = 5.23$ BM at 289.9°K and $\mu_{eff} = 0.79$ BM at 4.2°K, are consistent with the assumption of a $^5T_2 \rightleftharpoons {}^1A_1$ transition.[2] The sharp change of μ_{eff} at T_c suggests that the $^5T_2 \rightleftharpoons {}^1A_1$ transition is associated with a phase change. In contrast, the relevant magnetic properties of Fe(papt)$_2$ change gradually between 343°K and 83°K from those typical for the $^5T_2$ ground state to values somewhat in excess of those characteristic of the $^1A_1$ ground state, viz. $\mu_{eff} = 5.14$ BM at 343°K and $\mu_{eff} = 2.23$ BM at $\overline{83°K}$.[3] The temperature dependence of both $1/\chi_m^{corr}$ and μ_{eff} is displayed in Fig. 1b.

III. QUADRUPOLE SPLITTING AND ISOMER SHIFT OF ^{57}Fe MÖSSBAUER SPECTRA

The ^{57}Fe Mössbauer spectra[2] of Fe(4,7-(CH$_3$)$_2$-phen)$_2$(NCS)$_2$ between 298°K and 88°K are reproduced in Fig. 2. The values of the quadrupole splitting ΔE_Q and the isomer shift δ^{IS} for the single doublet observed at 298°K (viz. Table I) are sufficiently large to justify an assignment of the spectrum to the high-spin($^5T_2$)

Table I. ^{57}Fe Mössbauer effect data for Fe(4,7-(CH$_3$)$_2$-phen)$_2$(NCS)$_2$ at three typical temperatures.[a]

T(°K)	$^5T_2$		$^1A_1$		Area Fractions	
	ΔE_Q	δ^{IS} [b]	ΔE_Q	δ^{IS} [b]	I($^5T_2$)	I($^1A_1$)
298	2.59	+0.97	...	...	~0.98	~0.02
121.5	3.14	+0.98	0.47	+0.33	0.53	0.47
4.2	...	...	0.47	+0.32	0.04	0.96

[a]Quadrupole splittings ΔE_Q and isomer shifts δ^{IS} in mm sec^{-1}. The experimental uncertainty in ΔE_Q and δ^{IS} values is ± 0.02 and ± 0.03 mm sec^{-1}, respectively.
[b]Isomer shifts δ^{IS} are relative to natural iron at 298°K.

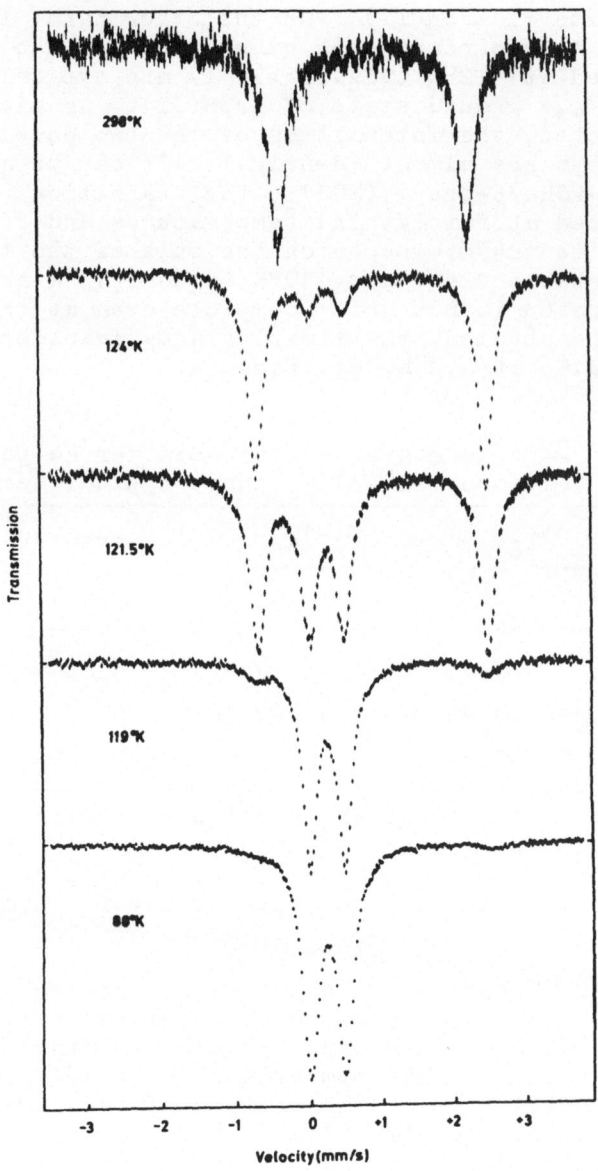

Fig. 2. ^{57}Fe Mössbauer spectra of Fe(4,7-(CH$_3$)$_2$-phen)$_2$-(NCS)$_2$ at a number of temperatures between 298 and 88°K. The three central spectra characterize the behavior close to T$_c$ = 121.5°K.

ground state of iron(II). On the other hand, the small
values of the corresponding quantities for the doublet
encountered at 4.2°K (viz. Table I) are typical for the
low-spin($^1A_1$) ground state of iron(II). As might have
been expected, the intensities of the two doublets at
T_c = 121.5°K are almost identical. It can be seen that,
in Fe(4,7-$(CH_3)_2$-phen)$_2$(NCS)$_2$, the transition $^5T_2 \rightleftharpoons {}^1A_1$
is initiated at fairly high temperatures and proceeds
gradually to 124°K, whereupon the bulk of the transition
follows between 124°K and 119°K (cf. Fig. 1a and Fig. 2).
The conversion is not quite complete even at the lowest
temperature studied, the limiting area fraction being
I($^1A_1$) = 0.96 at 4.2°K, cf. Table I.

Table II. ^{57}Fe Mössbauer effect data for Fe(papt)$_2$
           (absorber B[3]) at three typical temperatures.[a]

T(°K)	$^5T_2$		$^1A_1$		Area Fractions	
	ΔE_Q	δ^{IS} b	ΔE_Q	δ^{IS} b	I($^5T_2$)	I($^1A_1$)
343	2.03	+0.87	...	...	1.00	0
223	2.14	+0.90	1.54	+0.36	0.58	0.42
4.2	2.19	+0.90	1.56	+0.32	0.17	0.83

[a]Quadrupole splittings ΔE_Q and isomer shifts δ^{IS} in
  mm sec^{-1}. The experimental uncertainty in ΔE_Q and δ^{IS}
  values is ±0.04 and ±0.05 mm sec^{-1}, respectively.
[b]Isomer shifts δ^{IS} are relative to natural iron at 298°K.

     The corresponding ^{57}Fe Mössbauer spectra[3] of
Fe(papt)$_2$ between 305°K and 90°K are illustrated in Fig.3.
Again, the values of ΔE_Q and δ^{IS} for the single doublet
observed at the highest temperature studied, i.e. 343°K
(viz. Table II), are characteristic of the high-spin($^5T_2$)
ground state of iron(II). At 4.2°K, the intense doublet
is essentially due to the spectrum of the low-spin($^1A_1$)
ground state, there being a residual contribution from
the $^5T_2$ state of I($^5T_2$) = 0.17. In the $^5T_2 \rightleftharpoons {}^1A_1$
transition region (cf. Fig. 1b and Fig. 3), both ground
states are simultaneously present with a temperature
dependent site fraction n_{5T_2} of $^5T_2$ ground state
complexes. This is illustrated by Fig. 4 where both n_{5T_2}

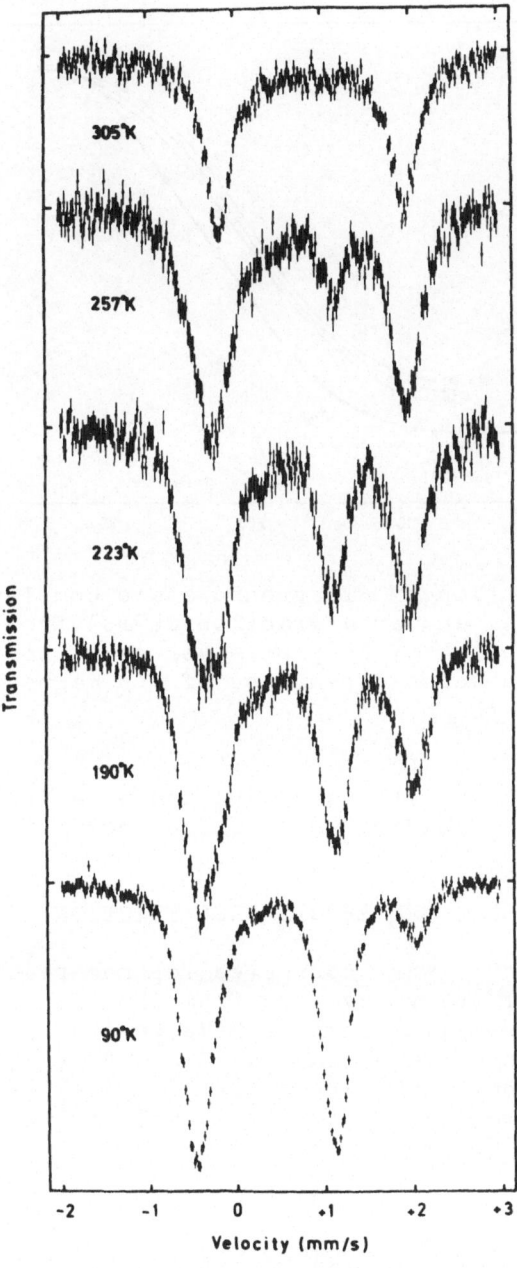

Fig. 3. ⁵⁷Fe Mössbauer spectra of Fe(papt)₂ at a number of temperatures between 305 and 90°K.

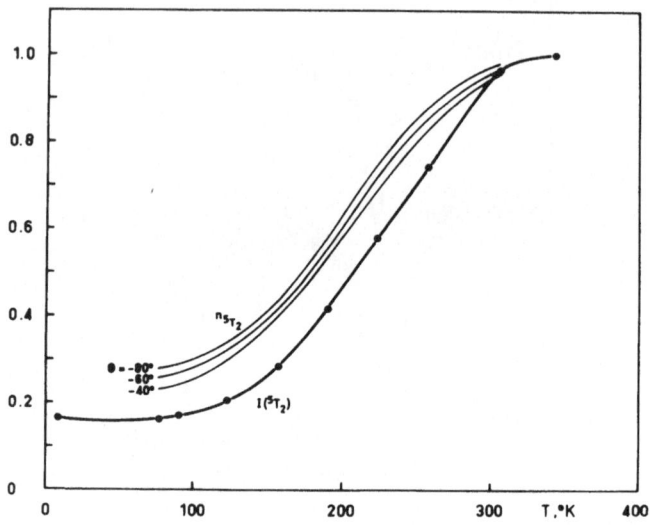

Fig. 4. Fraction of $^5T_2$ ground state complexes $n5T_2$ from magnetism and area fraction $I(^5T_2)$ from the Mössbauer spectra of Fe(papt)$_2$ versus temperature. Assumed Curie-Weiss dependence of $^5T_2$ magnetism indicated by values of Θ_p.

obtained from magnetic data and the area fraction $I(^5T_2)$ from Mössbauer spectra are plotted versus temperature.

IV. THE RECOIL-FREE FRACTION

The spectra were corrected for non-resonant background of the γ-rays and from the areas under the spectra Debye-Waller factors have been determined over the temperature range studied.[4,5] To this end, normalized individual areas $A(^5T_2)$ and $A(^1A_1)$ were extracted from the spectra ($v_{max} = \pm 4$ mm sec^{-1}), these being determined by

$$A(^5T_2) = f_S \Gamma \frac{\pi}{2} L(t5T_2)$$
$$A(^1A_1) = f_S \Gamma \frac{\pi}{2} L(t1A_1) \tag{1}$$

In Eq (1), f_S is the Debye-Waller factor of the source and the effective thickness t is given by

$$t_{5T_2} = \frac{1}{2} \sigma_0 \, w \, f_{5T_2} \, n_{5T_2}$$

$$t_{1A_1} = \frac{1}{2} \sigma_0 \, w \, f_{1A_1} \, n_{1A_1} \tag{2}$$

Here, the quantities f_{5T_2} and f_{1A_1} are the Debye-Waller factors for the $^5T_2$ and $^1A_1$ ground state, respectively, and n_{5T_2} and n_{1A_1} are the site fractions of molecules having the corresponding ground states. In addition, σ_0 is the resonant cross-section and w denotes the number of resonant nuclei per unit area of absorber. In the absorbers used containing 0.062 mg cm^{-2} ^{57}Fe (Fe(papt)$_2$) and 0.080 mg cm^{-2} ^{57}Fe (Fe(4,7-(CH$_3$)$_2$-phen)$_2$(NCS)$_2$), t is always smaller than 0.7 and the saturation function $L(t)$ is well approximated by

$$L(t) = t \, / \, (1 + 0.25 \, t) \tag{3}$$

Eq (1) - (3) would be applicable in the absence of field inhomogeneity if Γ is taken to be equal to Γ_0. The width of the observed lines in the spectrum is 0.25 mm sec^{-1} for Fe(4,7-(CH$_3$)$_2$-phen)$_2$(NCS)$_2$ and 0.29 mm sec^{-1} for Fe(papt)$_2$ at room temperature. The source (20 mCi ^{57}Co in copper) which is at room temperature during all measurements shown in Fig. 2 and Fig. 3 produces a line width of 0.23 mm sec^{-1} if a sodium nitroprusside absorber (0.1 mg cm^{-2} ^{57}Fe) is used. We assume that the absorber line shape results from the folding of a Lorentzian of width Γ_0 into a function which results from extra-nuclear statistical processes. If the latter function <u>is</u> Gaussian, the above procedure has to be modified as described by Lang.[4] In the present case, the corrections to the f values thus introduced amount to a maximum of about 3%. To obtain absolute f values, the quantity f_S was determined from measurements on thin absorbers of known Debye-Waller factors.

In the case of Fe(4,7-(CH$_3$)$_2$-phen)$_2$(NCS)$_2$, the contribution to the total area A_{total} of the $^1A_1$ ground state above T_c and that of the $^5T_2$ ground state below T_c are minor. Therefore, total Debye-Waller factors f_{total} have been determined between 298°K and 4.2°K,[3] <u>cf.</u> Fig. 5. It is evident that $-\ln f_{total}$ shows a well-defined discontinuity of ~17% at T_c. Moreover, its behavior neither above nor below T_c is reproduced by the Debye model, the Debye function with $\theta = 134$°K having been included in Fig. 5. In this case, it is reasonable

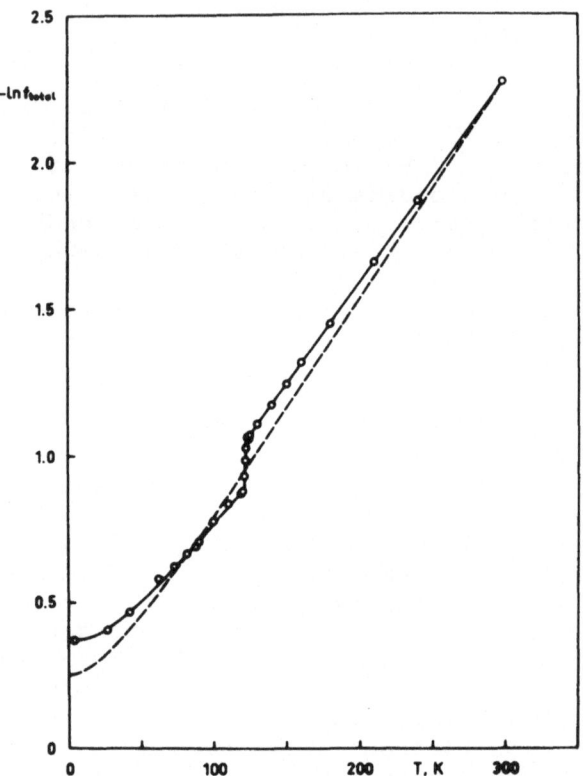

Fig. 5. Temperature dependence of $-\ln f_{total}$ from ^{57}Fe Mössbauer spectra of $Fe(4,7-(CH_3)_2-phen)_2(NCS)_2$. Broken curve gives $-\ln f$ calculated on the basis of the Debye model with $\Theta = 134^\circ K$. The transition temperature is $T_c = 121.5^\circ K$.

to assume

$$f_{5T_2} = f_{total} \quad \text{at } T > T_c$$
$$f_{1A_1} = f_{total} \quad \text{at } T < T_c \qquad (4)$$

and thus the discontinuity corresponds essentially to a change between f_{5T_2} and f_{1A_1}. If refinements are introduced to take account of the $^1A_1$ state above T_c and the $^5T_2$ state below T_c, the individual Debye-Waller

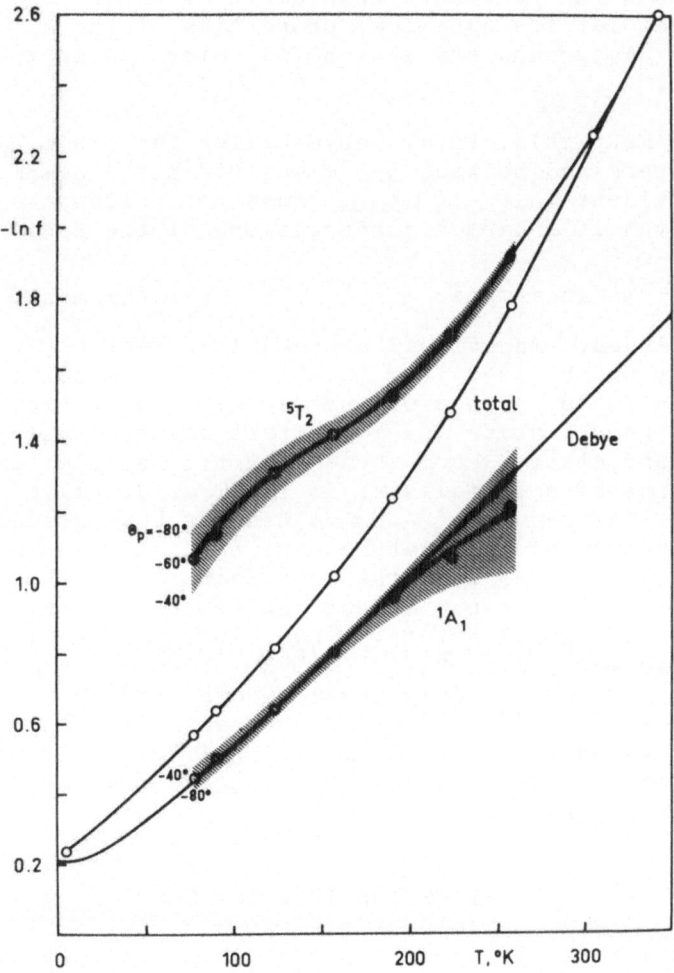

Fig. 6. Temperature dependence of $-\ln f_{5T_2}$ and $-\ln f_{1A_1}$ from ^{57}Fe Mössbauer spectra of Fe(papt)$_2$. Shaded areas indicate the range of $-\ln f_{5T_2}$ and $-\ln f_{1A_1}$ values which result subject to uncertainties in the Curie-Weiss law (Θ_p between -40 and -80°K). Curve marked 'total' refers to $-\ln f_{total}$, curve marked 'Debye' gives $-\ln f_{1A_1}$ as calculated from the Debye model using $\Theta_{1A_1} = 165$°K.

factors at T_c follow as $f_{5T_2} = 0.34$ and $f_{1A_1} = 0.41$.
The result then provides conclusive evidence for
different lattice dynamical properties at the sites of
$^5T_2$ and $^1A_1$ ground state iron(II) molecules in the
crystal.

In Fe(papt)$_2$, total Debye-Waller factors f_{total} have
been determined between 343°K and 4.2°K,[4] cf. Fig. 6.
It is evident that $-\ln f_{total}$ does not follow the Debye
model even if anharmonic corrections of the form

$$\ln f(T)_{anharmonic} = (1 + \epsilon T) \ln f(T)_{harmonic} \qquad (5)$$

are included. Individual Debye-Waller factors f_{5T_2} and
f_{1A_1} may be obtained from Eq (1) - (3) by assuming a
Curie-Weiss dependence of the magnetic susceptibility for
the $^5T_2$ ground state and a constant value of μ_{eff} for the
$^1A_1$ ground state. From these assumptions, the temperature
dependence of n_{5T_2} follows, in fact, as displayed in
Fig. 4. The resulting $-\ln f$ values for both ground states
as a function of temperature have been plotted in Fig. 6.
It should be noted that the uncertainties in the assumed
Curie-Weiss law do not affect the general behavior of
$-\ln f$ to any great extent. Thus $-\ln f_{1A_1}$ closely follows
the Debye model with $\Theta_{1A_1} = 165°K$, whereas the same
applies to $-\ln f_{5T_2}$ only above ~210°K and $\Theta_{5T_2} = 134°K$.
It should be noted that the Debye range in $-\ln f_{5T_2}$ is
characterized in that only an insignificant fraction of
molecules has been transformed into the $^1A_1$ ground state
(cf. Fig. 3 and Fig. 6).

V. HYPERFINE INTERACTIONS

Additional information concerning the molecular
geometry and the electronic structure in the $^5T_2$ and $^1A_1$
ground states may be obtained by a study of simultaneous
magnetic and electric hyperfine interactions. At 4.2°K,
hyperfine interactions can be studied in the $^1A_1$ ground
state only, since the fraction of molecules in the $^5T_2$
ground state is small, particularly so in the case of
Fe(4,7-(CH$_3$)$_2$-phen)$_2$(NCS)$_2$. In addition, the effective
magnetic field is given by

$$\underline{H}_{eff} = \underline{H}_{ext} + \underline{H}_n \qquad (6)$$

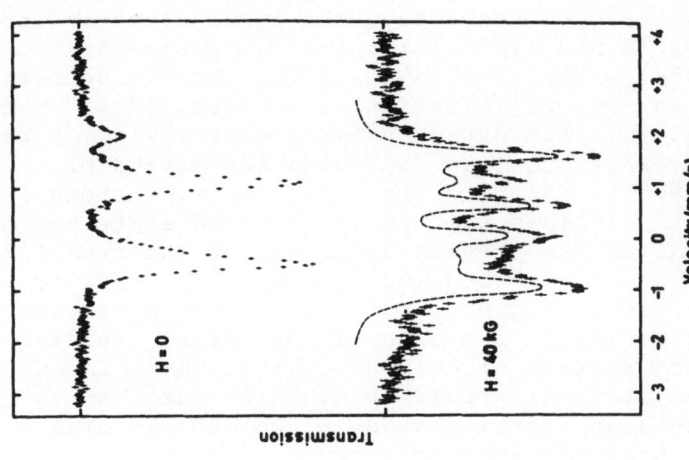

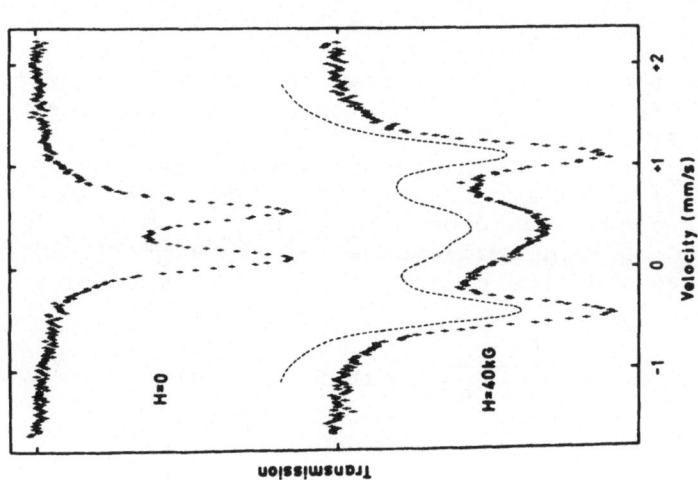

Fig. 7a and 7b. 57Fe Mössbauer hyperfine spectra of Fe(4,7-(CH₃)₂-phen)₂(NCS)₂ (left, a) and of Fe(papt)₂ (right, b) at 4.2°K: Hext = 0, upper spectra; Hext = 40 kG, lower spectra. Broken line gives the spectra calculated under the assumptions $V_{zz} > 0$, $\eta = 0.8$, $H_n = 0$ (a); $V_{zz} < 0$, $\eta = 0.7$, $H_n = 0$ (b).

where $\underline{H}_n$ is the internal magnetic field generated by $\underline{H}_{ext}$. In Fe(papt)$_2$ at 4.2°K then, the application of an external magnetic field of 40 kG distributes the small intensity of the $^5T_2$ spectrum over eight lines within about 7 mm sec^{-1} thus producing only a minor variation of the base line. Consequently, the apparent hyperfine spectra at 4.2°K are typical for the $^1A_1$ ground state. In Fe(4,7-(CH$_3$)$_2$-phen)$_2$(NCS)$_2$ at 4.2°K, the $^1A_1$ doublet is split by an external (longitudinal) field of 40 kG as shown in Fig. 7a. The experimental spectrum[6] may be reproduced by a calculation following the method of Collins and Travis,[7] and the result is also shown in the figure. It follows that the $^1A_1$ ground state may be characterized, in the present compound, by the hyperfine parameters $V_{zz} > 0$, $\eta \sim 0.8$ and $H_n = 0$. Here, V_{zz} is the electric field gradient, η the asymmetry parameter, and H_n the strength of the internal hyperfine field. The hyperfine spectrum of Fe(papt)$_2$ at 4.2°K[8] in an external (longitudinal) field of 40 kG is displyed in Fig. 7b. The best approximation to the experimental spectrum has been achieved again by a Collins and Travis[7]-type calculation whereby, in this case, it has been assumed that $V_{zz} < 0$, $\eta \sim 0.7$ and $H_n = 0$.

For a study of hyperfine interactions in the $^5T_2$ ground state, the temperature has to be raised significantly as is clearly evident from Fig. 2 and Fig. 3. For the $^5T_2$ ground state, an internal magnetic field $\underline{H}_n$ is expected and indeed the splitting of each line is too small (viz. 0.44 and 0.78 mm sec^{-1} in Fe(4,7-(CH$_3$)$_2$-phen)$_2$(NCS)$_2$) compared with the normal splitting of $\sim$1.0 mm sec^{-1} at 40 kG. Under the experimental conditions chosen (i.e. H_{ext} = 40 kG, T = 163°K and 180°K), small values of the magnetization may be assumed and thus $\underline{H}_n$ may be expressed according to[9]

$$H_{ni} = \frac{\chi_i H_{ni}^{(0)}}{N g_i \mu_B S} H_{ext,i} \qquad i = x,y,z \qquad (7)$$

where χ is the susceptibility tensor, $H_n^{(0)}$ is the saturation value of the internal magnetic field, S=2 for iron(II), and where the remaining quantities have their usual meaning. In Fe(4,7-(CH$_3$)$_2$-phen)$_2$(NCS)$_2$, a pure $^5T_2$ spectrum is observed in zero field at a temperature of 163°K. Both the H_{ext} = 0 spectrum and the spectrum measured at 40 kG[6] are shown in Fig. 8a. Within the

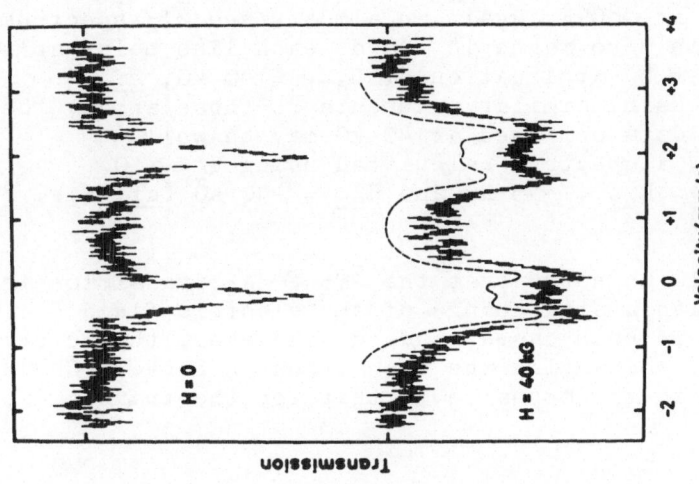

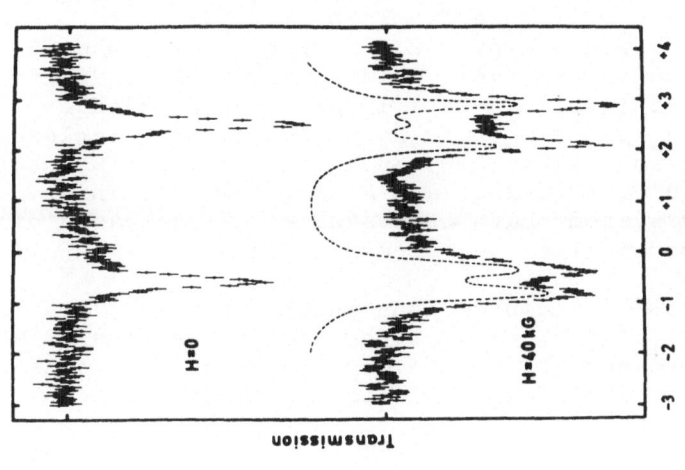

Fig. 8a and 8b. ^{57}Fe Mössbauer hyperfine spectra of Fe(4,7-(CH$_3$)$_2$-phen)$_2$(NCS)$_2$ at 163°K (left, a) and of Fe(papt)$_2$ at 180°K (right, b): H$_{ext}$ = 0, upper spectrum; H$_{ext}$ = 40 kG, lower spectrum. Broken line gives the spectra calculated under the assumptions $V_{zz} < 0$, $\eta = 0$, H$_{nx}$ = H$_{ny}$ = -2 kG, H$_{nz}$ = -23 kG (a); $V_{zz} > 0$, $\eta = 0.75$, H$_{nx}$ = H$_{ny}$ = -15 kG, H$_{nz}$ = -10 kG (b).

above approximation, the experimental spectrum is well
reproduced if $V_{zz} < 0$, $\eta \sim 0$, $H_{nx} = H_{ny} = -2$ kG and
$H_{nz} = -23$ kG are assumed (viz. Fig. 8a, broken curve).
In Fe(papt)$_2$ at 180°K, again an almost pure $^5T_2$ spectrum
is observed in zero magnetic field, each line being split
into a doublet by application of $H_{ext} = 40$ kG, cf.
Fig. 8b.[8] Using similar arguments to those above, the
experimental data obtained at 40 kG may be well
reproduced by a spectrum calculated using $V_{zz} > 0$,
$\eta \sim 0.7$, $H_{nx} = H_{ny} = -15$ kG and $H_{nz} = -10$ kG (cf. Fig. 8b,
broken curve).

It should be noted that the $^5T_2 \rightleftharpoons {}^1A_1$ transition is
associated with a sign change of the electric field
gradient V_{zz} in both cases studied. However, the actual
sign change accompanying the transition in Fe(4,7-(CH$_3$)$_2$-
phen)$_2$(NCS)$_2$ is the opposite to that for the transition
in Fe(papt)$_2$.

VI. DISCUSSION

The behavior discussed above seems to be typical for
$^5T_2 \rightleftharpoons {}^1A_1$ transitions in solid complexes of the [Fe^{II}-N$_6$]
type. Results refering to sharp $^5T_2 \rightleftharpoons {}^1A_1$ transitions
which are similar to those obtained in Fe(4,7-(CH$_3$)$_2$-
phen)$_2$(NCS)$_2$ are available for Fe(bipy)$_2$(NCS)$_2$ and
Fe(phen)$_2$(NCS)$_2$ where bipy = 2,2 -bipyridyl and phen =
1,10-phenanthroline.[10-12] Gradual $^5T_2 \rightleftharpoons {}^1A_1$ transitions
have been investigated in detail for Fe(pythiaz)$_2$X$_2$,[13]
Fe(2-CH$_3$-phen)$_3$X$_2$,[14,15] and Fe(paptH)$_2$X$_2$[12] complexes.
Here, pythiaz is 2,4-bis(2-pyridyl)thiazole, paptH denotes
the protonated papt ligand, and X is a suitable anion.
Various other systems have been studied on the basis of
magnetic susceptibility measurements alone.

Those $^5T_2 \rightleftharpoons {}^1A_1$ conversions which are associated with
a sharp change of the relevant physical properties at a
specific temperature T_c may be described best in terms of
a first order transition. This is supported, e.g., by
the change in entropy encountered in the parent complex,
Fe(phen)$_2$(NCS)$_2$, at $T_c = 176.29$°K. In fact, a distinct
anomaly in the heat capacity has been observed[16,17]
at T_c, and $\Delta H = 8.61$ kJ mol^{-1}, $\Delta S = 48.78$ JK^{-1}mol^{-1} were
obtained from a detailed study of C_p. Since the entropy

change due to spin reorientation is determined by

$$\Delta S_{spin} \sim R \ln 5 = 13.38 \text{ JK}^{-1}\text{mol}^{-1} \qquad (8)$$

the excess entropy has been assumed[16] to be due to enhanced phonon excitation in the $^5T_2$ ground state.

However, since $\Delta H = T_c \Delta S$, the total entropy change is

$$\Delta S = \Delta S_{config} + \Delta S_{vib} + \Delta S_{el} \qquad (9)$$

and positive if the temperature is increased through T_c. In Eq (9), the first and the second terms are the configurational and the vibrational entropy, respectively, of the nuclei including core electrons, and ΔS_{el} is the entropy due to the valence electrons, $\Delta S_{el} \sim \Delta S_{spin}$. It should be noted that ΔS_{config} is not necessarily small, since a change of Fe-N bond distances may be expected. This is due to the fact that, in the $^5T_2 \rightleftharpoons {}^1A_1$ transition, two e electrons are transformed into t_2 electrons. Additional changes in the geometry of the $[Fe^{II}-N_6]$ complex may occur as exemplified by the X-ray structural study on $Fe(bipy)_2(NCS)_2$.[18] Provided that the results obtained for $Fe(4,7-(CH_3)_2-phen)_2(NCS)_2$ are typical, the decrease of f_{total} at T_c indicates that $\Delta S_{vib} > 0$ but small ($\Delta S_{vib} \sim 2.3$ JK^{-1}mol^{-1} on the basis of the Debye model). It follows, in conjunction with Eq (8) and Eq (9), that

$$\Delta S_{config} \sim 33 \text{ JK}^{-1}\text{mol}^{-1} \qquad (10)$$

is the driving force of the first-order transition. This conclusion is supported by the near-equality of the value of ΔS quoted above and the value $\Delta S = 47.71$ JK^{-1}mol^{-1} found[1] for the $^5T_2 \rightleftharpoons {}^1A_1$ equilibrium in solution. In addition, the mass of the ^{57}Fe Mössbauer atom is much greater than the average mass of the other atoms in the complexes, and thus the localized (i.e. intramolecular) vibrations will not contribute significantly to the recoil-free fraction. The behavior of f_{total} then suggests a freezing out of lattice vibrations due to increased vibrational frequencies above the transition temperature, although the effect may be small. Consequently, the transition will be slightly phonon assisted.

In $Fe(bipy)_2(NCS)_2$, the overall symmetry of the $^1A_1$ ground state is higher than that of the $^5T_2$ ground state. It has been suggested[10] that the energy separation between the $^5T_2$ and $^1A_1$ states may be overcome by the low-symmetry splitting (or shift, for that matter) of the $^5T_2$ state. This process would then be similar to cooperative distortions encountered in some electron-phonon coupled Jahn-Teller systems. The sign change of V_{zz} accompanying the $^5T_2 \rightleftharpoons {}^1A_1$ transition in the complexe studied above may then be taken as a consequence of a configurational change analogous to that in $Fe(bipy)_2(NCS)_2$.

The gradual $^5T_2 \rightleftharpoons {}^1A_1$ transition seems to be the basic phenomenon, although it is sometimes interrupted and dominated by the discontinuous change discussed above Unfortunately, on the time scale of the Mössbauer effect (precession time of the ^{57}Fe nucleus in the range 10^{-7} to 10^{-9} sec), the same spectra would be observed whether a dynamical transition typified by a rate constant $k \lesssim 10^7$ sec^{-1} (similar to that observed in solution[1]) or a static $^5T_2 \rightleftharpoons {}^1A_1$ conversion is involved. The present authors prefer the latter interpretation as a working hypothesis. Indirect evidence in support of this assumption is the anomalous behavior of the Debye-Waller factor as well as the observation of a residual fraction of $^5T_2$ ground state molecules at cryogenic temperatures. In fact, this residual fraction is a characteristic feature of complexes displaying $^5T_2 \rightleftharpoons {}^1A_1$ transitions and it increases in some systems up to 63%.[15] Also at variance with the assumption of a dynamical transition ("spin equilibrium") is the observation that the plot of $\ln n_{5T_2}/n_{1A_1}$ versus $1/T$ does not produce a straight line. Obviously, the recently reported[19,20] pressure induced $^5T_2 \rightleftharpoons {}^1A_1$ transitions may be understood along similar lines.

Acknowledgement. The authors appreciate discussions and comments on the manuscript by H. A. Goodwin.

REFERENCES

1. J. K. Beattie, N. Sutin, D. H. Turner, and G. W. Flynn, J. Am. Chem. Soc. 95, 2052 (1973).
2. E. König, G. Ritter, and B. Kanellakopulos, J. Phys. C in the press.

3. E. König, G. Ritter, and H. A. Goodwin, to be
    published.
4. G. Lang, Nucl. Instr. Meth. 24, 425 (1963).
5. G. A. Bykov and Pham Zuy Hien, Soviet Phys. JETP 16,
    646 (1963).
6. E. König, G. Ritter, and R. Zimmermann, Chem. Phys.
    Lett., in the press.
7. R. L. Collins and J. C. Travis, Mössbauer Effect
    Methodology, I. J. Gruvermann Ed., 3, 123 (1967).
8. E. König, G. Ritter, R. Zimmermann, and H. A.
    Goodwin, J. Chem. Phys., in the press.
9. C. E. Johnson, Proc. Phys. Soc. 92, 748 (1967).
10. E. König and G. Ritter, Phys. Lett. 43A, 488 (1973).
    Cf. also E. König, K. Madeja, and K. J. Watson,
    J. Am. Chem. Soc. 90, 1146 (1968).
11. I. Dezsi, B. Molnar, T. Tarnoczi, and K. Tompa, J.
    Inorg. Nucl. Chem. 29, 2486 (1967). Cf. also
    E. König and K. Madeja, Inorg. Chem. 6, 48 (1967).
12. E. König and G. Ritter, unpublished results.
13. E. König, G. Ritter, and H. A. Goodwin, Chem. Phys.
    1, 17 (1973).
14. E. König, G. Ritter, H. Spiering, S. Kremer, K.
    Madeja, and A. Rosenkranz, J. Chem. Phys. 56, 3139
    (1972).
15. E. König, G. Ritter, B. Braunecker, K. Madeja, H. A.
    Goodwin, and F. E. Smith, Ber. Bunsenges. Phys. Chem.
    76, 393 (1972).
16. M. Sorai and S. Seki, J. Phys. Soc. Japan 33, 575
    (1972).
17. M. Sorai and S. Seki, J. Phys. Chem. Solids, in the
    press.
18. E. König and K. J. Watson, Chem. Phys. Lett. 6, 457
    (1970).
19. D. C. Fisher and H. G. Drickamer, J. Chem. Phys. 54,
    4825 (1971).
20. C. B. Bargeron and H. G. Drickamer, J. Chem. Phys.
    55, 3471 (1971).

Self-Absorption Mössbauer Investigation of Neutron-Activated
Krypton

John B. Brown, Jr.

Battelle Columbus Laboratories - Durham
Operations
3333 Chapel Hill Boulevard, Durham, N. C.

INTRODUCTION

The Debye-Waller factor F is related to the intensity
of x-rays and cold neutrons recoillessly or elastically
scattered from crystals. Similarly, the Mössbauer factor,
f, gives the fraction of gamma rays emitted or absorbed
with no recoil in a Mössbauer experiment. Here, the term
recoilless interaction means that recoil energy does not
go to the degrees of freedom of the lattice system but
only to the recoil of the affected aggregate of atoms or
crystal as a stiff, rigid unit. In the study presented
here, the recoilless fraction, f, is investigated for solid
krypton at 4.2° K. Specifically, f is determined by measuring the degree of self-abosrption of 9.3 KeV resonant
and non-resonant radiation in a solid disk of neutron activated krypton as a function of disk thickness. Recoilless
fraction (f) dependent resonance and non-resonance source
photons within the Kr source disk are affected by high
cross section f-dependent absorption and f^2-dependent
scattering interactions within the source. This strong
f-dependence of source strength and interaction probability
thus manifests itself in the measurement of non-resonant
and resonant photon self-absorption. Self-absorption is
determined by measuring the radiation intensity of 9.3 KeV
photons escaping the source disk.

The measurement of Mössbauer recoilless fraction in solid krypton at LHe temperature by a self-absorption scheme has several nice features.

Solid krypton, which possesses a Mössbauer isotope Kr^{83} (11.55% abundant), is a good material for studying the recoilless fraction since it is nonmagnetic and cubic in structure. The analysis which assumes that there are no solid-state effects perturbing the natural Lorentzian shape or width of decay and absorption peaks is expected to hold.

With the utilization of self-absorption data to determine the Mössbauer recoilless fraction f in Kr ice at 4.2° K, a second independent experimental technique to determine f is evaluated.

The internal conversion coefficient, α, has recently been remeasured by Kolk (1,2) and found greater by 64% than the heretofore published value as measured by Ruby, et al. (3). The parameter α is introduced into the Mössbauer absorption cross section at resonance by the multiplication factor $(1 + \alpha)^{-1}$ and into the resonant scattering cross section by the factor $(1 + \alpha)^{-2}$. After reducing the resonant cross section as established by use of the new value of α, recent investigators were able to bring into agreement published experimental values of f and new theory. However, considering the newness, the large percentage change, and the lack of independent verification of the recent measurement of α, results obtained from self-absorption techniques indicating the value of α are of interest as an independent check.

LITERATURE REVIEW

Soon after lifetime measurements (3) of the 9.3 KeV state in Kr^{83}, Hazony, et al. (4), measured the Mössbauer effect in this isotope; firstly in frozen krypton and secondly krypton trapped in a clathrate compound. In the frozen krypton experiment, a polythene absorber bag filled with natural krypton gas was strapped onto a metal strip dipping into liquid air. Irradiation of similar source bag for 10 minutes at 10^{13} n-cm^{-2}-sec^{-1} produced, via the Kr^{82} (n,γ) Kr^{83m} reaction, an adequate strength of Kr^{83} in its 114 minute isomeric 42 KeV state. Utilizing a sinu-

soidal drive and modulated multichannel analyzer, a small
positive on-off effect was found with essentially unknown
absorber thickness.

Ruby and Selig (5) continued the Mössbauer study of
Kr^{83} by utilizing the effect to help confirm that KrF_2 is
indeed closely analogous to XeF_2. Experimentally, the
Mössbauer experiment was a typical separated source and
absorber scheme. Ruby and Selig point out that low specific
activities from neutron activation of natural krypton are
an incentive for the formation of thick Mössbauer sources
for adequate counting statistics. However, thick krypton
sources manifest themselves in experimental results by
broadening the experimental peak (line width) observed in
conventional Mössbauer spectra. Absorber reactions thus
escape proper identification from Mössbauer spectra if
source thickness effects are not properly accounted for.

Ruby and Selig (5) reported a value of 0.33 ± 0.33
mm/sec for the line width of solid krypton (both source and
absorber). Since the minimum experimental line width 2Γ
(for a "natural" source and absorber, both thin) is calculated to be 0.20 mm/sec, the extra width experienced by
Ruby and Selig is probably due to the effects of the black-
ness or self-absorption of the source and absorber. Ruby
and Selig made no mention of recoil-free fraction for their
krypton ices.

Pasternak, et al. (6) reported results of measurements
performed with solid sources. The absorber of solid krypton
was produced as a thin layer by spraying gas through a
cryostat port onto a cold aluminum backing of $3mg/cm^2$
thickness. The absorber thickness was measured using
attenuation of the 6.3 KeV x-rays of Fe^{57}. The reported
line width was 0.82 mm/sec for a krypton ice absorber at
32° K. By measuring the effect of temperature variation
on the absorption peak and assuming a Debye model for the
solid krypton phonon spectrum, Pasternak calculated the
Debye temperature, Θ_D, to be 37° K.

The next year Pasternak and Sonnino (7) studied the
Mössbauer effect of Kr^{83} in bromide and bromate crystals.
Absorbers used in this investigation were solid krypton
and β-hydroquinone clathrates. Again the solid krypton
absorber was produced by slow injection of gas onto an
aluminum backing of 3 mg/cm² thickness kept at 22° K.

Under optimum conditions the deposition efficiency was
reported to be 80%. The thickness of the krypton layer
again was measured by the absorption of 6.3 KeV x-rays of
a Co^{57} source. From experiments with "thick" solid krypton
absorbers, Pasternak and Sonnino calculated the absolute
"effective" recoilless fraction, f_{eff} , of the krypton
clathrate source. This effective fraction, f_{eff} , is less
than the absolute f_s, due to the fact that the source
exhibits a self-resonant absorption. The fraction, f_{eff} is
a function of the effective source-resonant thickness, and
the electronic absorption.

Bukshpan, et al. (8) produced new sources from Mössbauer·
effect experiments in Kr^{83} by decay of Se^{83} via Br^{83} to Kr^{83m}
These sources in the form of ZnSe, PbSe and SnSe produced
lines of natural line width against solid Kr absorbers. A
relative shift of the Mössbauer spectrum peak between ZnSe
and PbSe in the clathrate experiment was discussed in terms
of differences in the square of the zero angular momentum
electronic wave function. No mention was made of recoil-
free fractions in solid krypton.

Gilbert (9) and Gilbert and Violet (10) performed a
theoretical and experimental investigation of the lattice
dynamics of solid krypton; in both cases determining the
atomic mean square displacement $\langle x^2 \rangle$ as a function of
temperature. In the experimental work, $\langle x^2 \rangle$ as a function
of temperature was obtained by using a conventional Mössbauer effect experiment to measure the recoilless fraction
in solid Kr over the temperature range 5 to 85° K. In
theoretical work, $\langle x^2 \rangle$ as a function of temperature was
calculated from the spatial distribution function, itself
determined by a Monte Carlo calculation using the Lennard-
Jones potential. The experimental recoilless fractions
determined were significantly lower than predicted from
the nearly-identical Kr frequency spectra of Brown and
Horton (11) or Jelinek (12), Figure 1(a). Experimental
recoilless fractions (f values) are seen to be lower than
the theoretical values. Corresponding mean square dis-
placements were reported to be 1.5-2.5 times theoretical
values; effective Debye temperatures were 20-60% below
theoretical values. As presented earlier, Pasternak, et al.
(6) quoted a Debye temperature of 37° K for a Kr temperature
of approximately 50° K. This agreed with Gilbert and Violet
(10). It was surmised by Gilbert and Violet that the low
f's could be caused by an unusually high vacancy concentra-

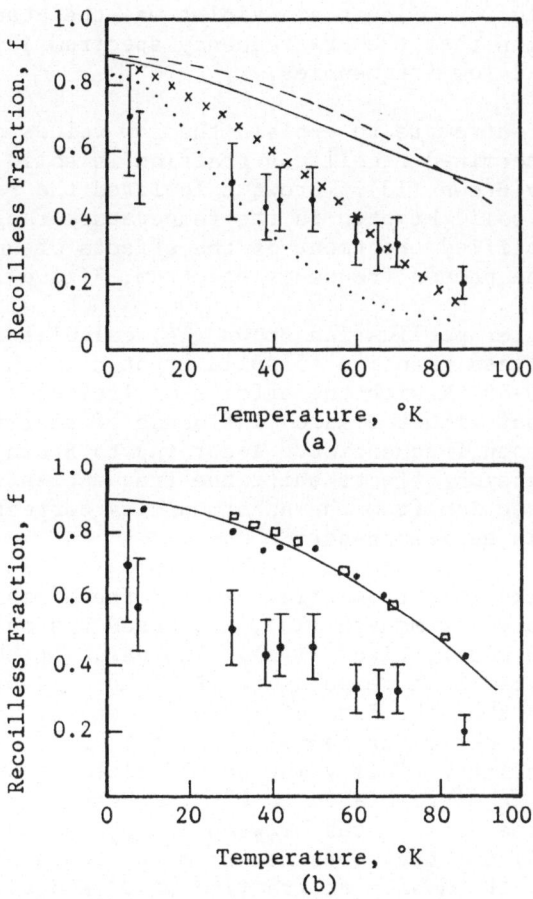

Figure 1. Recoilless fraction as a function of temperature;
(a) experiment (error barred) and calculation (dashed curve)
of Gilbert-Violet (10), calculation of Brown (13) including
anharmonicity (x-curve) Brown, (14) including anharmonic and
thermal expansion frequency shifts (solid curve), small dots
indicate Debye-model results of Mahesh (15); (b) Kolk (1)
corrected Gilbert-Violet data (small dots) compared with
original Gilbert-Violet (11) values (error barred), Kolk (1)
theoretical calculation (solid curve) and Windecker (16)
calculated values (squares).

tion in their source and absorber ices, which in turn might
be associated with their method of deposition. Another
interpretation by Gilbert and Violet was the abandonment of
the assumption that the Kr frequency spectrum is necessarily
Debye-like at low frequencies.

Further attempts to explain the low values of experimentally determined recoilless fraction in solid krypton
were made by Brown (13). Brown calculated the recoilless
fraction of solid krypton in the temperature range 0-85° K,
using a simplified treatment of the effects of anharmonicity upon the phonon frequency spectrum, Figure 1(a).

In a later publication Brown (14) calculated the Mössbauer recoilless fraction of solid krypton in the temperature range 0-85° K with the effects of lattice anharmonicity
as before, but with the added influence of thermal expansion
upon the phonon frequencies. According to Brown, the
thermal expansion effects shift the frequencies in the
opposite direction from the anharmonicity correction included in his earlier paper.

For comparison Brown (14) plotted the Mössbauer fraction calculated by Mahesh (15), who based his calculations
on the Debye model, Figure 1(a). The value of the Mössbauer characteristic temperature Θ_m (Θ_0) used by Mahesh
was obtained from the data of Pasternak et al. (6), who
estimated the Mössbauer temperature to be 37° K as was
discussed earlier. This value was obtained at 50° K, however, and when extrapolated to 0° K, Mahesh found $\Theta_m=42°$ K.
Because of the differences between the phonon spectrum used
by Brown (14) and that obtained from the Debye model at low
frequencies, the Mössbauer fraction at 0° K obtained by
Mahesh (15) is in better agreement with the Gilbert-Violet
(10) data. The Mahesh curve falls below the Gilbert-Violet
data for higher temperature as Mahesh's shift includes only
thermal expansion effects and consequently overestimates
the anharmonic correction. Conversely, Brown's (14) curve
falls above the Gilbert-Violet data.

The recoilless fraction depends on the mean square
displacement of the atoms in the lattice in the same way
as the intensities of x-ray Bragg reflections. Integrated
intensity measurements of x-ray Bragg reflections as a

function of the temperature were recently performed on
solid krypton by Windecker (16). His results agree with
theory if the anharmonicity is taken into account, but they
disagree with the Mössbauer data of Gilbert-Violet, Figure
1(b).

This disagreement of the results of Gilbert-Violet
with the experimental results of Windecker (16), indicated
by Kolk (1), suggests that there may be an error in the
evaluation of the absolute f-value from the Mössbauer measurements. To derive the absolute value of the recoilless
fraction from Mössbauer absorption measurements, the nuclear
resonance cross section σ_o is needed (17);

$$\sigma_o = \frac{1}{2\pi} \cdot \frac{hc}{E_\gamma} \cdot \frac{1}{1+\alpha_{10}} \cdot \frac{2I_1+1}{2I_0+1}$$

Here E_γ is the gamma ray energy α_{10} the total internal conversion coefficient of the gamma transition and I_1 and I_0
are the spins of the excited and ground state of the nucleus.
Until very recently, a value (18) $\alpha = 11 \pm 2$ has been used
for the total conversion coefficient in the equation for σ_o.
However, in the most recent work of Kolk (1,2) the experimental f-values of Gilbert-Violet using, however, the Kolk-
measured values of the conversion coefficient $\alpha_{10} = 18.3 \pm$
0.06 and the gamma transition energy $E_\gamma = 9.4$ KeV, have
been brought into agreement with those as expected from
the spectra of Brown and Horton (11). In Figure 1(b), the
original and corrected f-values are shown. These values
of α_{10} and E_γ as compared to the customarily accepted ones
(11 $\pm$ 2 and 9.3 KeV, respectively) are likely to be the
subject of further study in view of this difference.

With the exception of the present work, all experiments involving Mössbauer effect in Kr^{83} were performed in
transmission geometry. That the resonant line space
traced out in this transmission experiment is dependent on
source self-absorption coupled with the need to measure
such lines for data interpretation, makes a consideration
of source self-absorption effects of interest. Margulies
and Ehrman (19) investigated the transmission of resonant
gamma radiation emitted from a source of finite thickness,
and passing through a finite resonance absorber.

Margulies and Ehrman obtained the results:

$$T(\bar{E}) = \left[(1-f) T \binom{\text{uniform}}{\text{non-res}} + f \left(1 - \frac{T_s}{4} \right) \right] - f \frac{T_A}{2} \frac{1}{1+(\bar{E}/\Gamma)^2}$$

The bracketed term in this equation represents the transmission when $\bar{E}$, the relative Doppler shift between source and absorber, is large. This asymptotic value is less than unity because of source self-absorption. The second term represents the dip in the transmission due to resonance absorption in the external absorber. In the case of thin source and absorber, the transmitted line has a Breit-Wigner shape and the apparent width Γ_{obs} is twice that of either the emission or absorption spectrum.

For the general case Margulies and Ehrman could find no analytical evaluation. A computer solution, however, resulted in the transmitted spectrum $T(\bar{E})$ to be broadened beyond a value equal to the sum of individual source and absorber spectra.

Physically speaking, the uniform source atom distribution case evaluated by Margulies and Ehrman is identical to a neutron activated solid krypton source. However, by virtue of self-absorption and self-resonant scatter, a redistribution of emitting nuclear states (source atoms) may exist to the degree that the above conventional source thickness correction scheme is not sufficient for correct interpretation of conventional Mössbauer data.

THEORETICAL BASIS FOR MÖSSBAUER SELF-ABSORPTION EXPERIMENT

The experiment described here utilizes photon self-absorption and self-scatter characteristics of neutron-activated, solid krypton for determination of Mössbauer recoilless fraction f at liquid helium temperature. It is expected that these 9.3 KeV resonant and non-resonant quanta emanations from the source disk will be dependent on source-disk thickness in a different way than normally expected with only electronic scattering and absorption (Compton and Photoelectric interactions). Therefore, a measurement of the intensity as a function of energy characteristics of quanta emanations from such a disk, compared with theoretical predictions, is expected to give a measure of the number of resonant or recoilless interactions occurring within the source disk.

Two mathematical formulations describing the effects of resonant self-absorption are outlined. In the first formulation the distribution of excited Kr^{83} nuclear states is determined throughout the source disk, and from the known excited-state distribution (source-atom distribution), the photon intensities emanating from the source are determined. The second formulation utilizes the time independent Boltzmann equation to describe the resonant and non-resonant photon fluxes as a function of position, energy, and direction of motion.

Source Atom Distribution Scheme

Referring to the decay scheme and definitions given in Figure 2 and the creation and destruction of Mössbauer source nuclei in differential volumes depicted in Figure 3, the balance for the time rate of change for the concentration of Mössbauer source nuclei per cm^3, m_1, in dx at position x and time t is

$$\frac{\partial m_1(x,t)dx}{\partial t} = \lambda_2 P(x,t)dx - \lambda_1 m_1(x,t)dx + m_0 \overline{\sigma_R}(x)\phi(x' \to x,t)dx$$

$$- m_1 \overline{\sigma_R}(x)\phi(x' \to x, t)dx \; ;$$

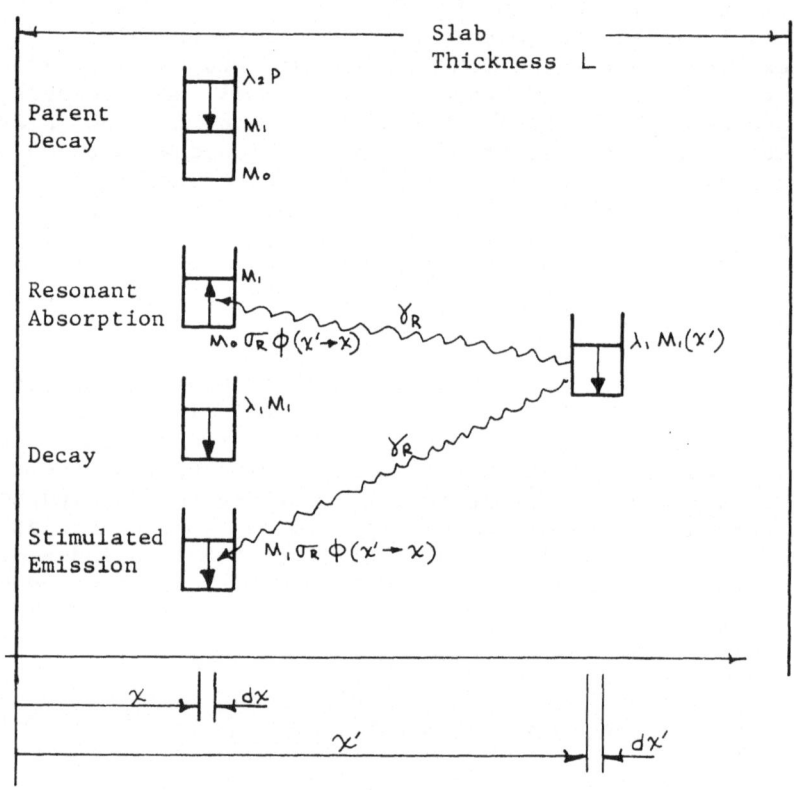

Figure 2. Schematic representation of the creation and destruction of Mössbauer source nuclei Kr^{83}, M_1 at slab positions x and x'; P(x,t) is the concentration of neutron activated Kr^{83m} nuclei at x per cm^3, λ_2 is the decay constant of Kr^{83m}, $M_1(x,t)$ is the concentration of Kr nuclei/cm^3 at x excited to the Mössbauer level, λ_1 is the decay constant of the Mössbauer nuclear level.

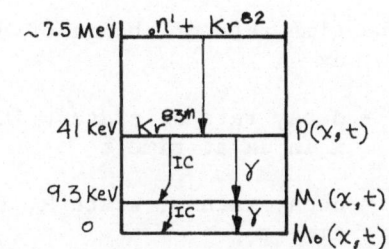

Kr^{83} nuclear level designations

$$P \xrightarrow{\lambda_2} M_1 \Bigg\langle \begin{matrix} \lambda_1 \left(\dfrac{\alpha_{10}}{1+\alpha_{10}}\right) \rightarrow M_0 + e^- \\[6pt] \lambda_1 \left(\dfrac{1-f}{1+\alpha_{10}}\right) \rightarrow M_0 + \gamma_{NR} \\[6pt] \lambda_1 \left(\dfrac{f}{1+\alpha_{10}}\right) \rightarrow M_0 + \gamma_R \end{matrix}$$

Decay options

$P(x,t)$ = # Kr^{83m} nuclei/cm^3 at x and t in state 2

$M_1(x,t)$ = # Kr^{83} nuclei/cm^3 at x and t that exist
           in the 9.3 KeV nuclear energy state

$M_0(x,t)$ = # Kr^{83} nuclei/cm^3 in the ground state

λ = decay constant for decay of state 1 or 2

α = total internal conversion coefficient

f = recoilless fraction

e^- = internal conversion electron

γ_{NR} = 9.3 KeV uncollided gamma ray emitted from
the Kr^{83} nucleus <u>with</u> atom recoil

γ_R = 9.3 KeV uncollided gamma ray emitted from
the Kr^{83} nucleus <u>without</u> atom recoil,

Figure 3. Schematic representation of nuclear reaction
Kr^{82} (n, γ) Kr^{83} with definition of terms.

where

$$\frac{\partial m_1(x,t)dx}{\partial t} = \text{ time rate of change of } M_1 (x,t) \text{ at x in dx}$$

$$\lambda_2 P(x,t)dx = \text{ decay rate of state P into state } M_1 \text{ at x in dx at time t}$$

$$\lambda_1 m_1(x,t)dx = \text{ decay rate of state } M_1 \text{ in dx at x at time t}$$

$$M_0 \, \overline{\sigma}_R \, \varphi(x',x,t)dxdx' = \text{ the nuclear absorption rate of } \gamma_R \text{ by } M_0 \text{ in dx at x as a result of } \gamma_R \text{ born in dx' at x'}$$

$$m_1 \overline{\sigma}_R \, \varphi(x',x,t)dxdx' = \text{ the stimulated emission rate of } \gamma_R(x) \text{ by interaction of excited nucleus } M_1 (x) \text{ with } \gamma_R(x' \ x)$$

and

$$\varphi(x' \rightarrow x,t) = \frac{\lambda_1 f}{1+\alpha_{10}} \int_{x'} dx' \, m_1(x',t) \, K(x' \rightarrow x) \quad ;$$

where $K(x' \, x)$ is the probability that a γ_R born at x' in dx' will travel to x in dx without electronic or nuclear interaction. Since $M_1 \ll M_0$ the time rate of change of M_1 reduces to:

$$\frac{\partial m_1(x,t)}{\partial t} = \lambda_2 P(x,t) - \lambda_1 m_1(x,t) + \frac{\Sigma_R f \lambda_1}{1+\alpha_{10}} \int_{x'} dx' \, m_1(x',t) \, K(x' \rightarrow x).$$

By assuming $\Sigma_R(x) \neq f(x)$, $P(x,t) \neq f(x)$, $m_1(x,t) = m_1(x)m_1(t)$ and that $P(x,t) = P_0 m_1(t)$; the resulting equation describing the distribution of atoms in the excited Mössbauer state M_1, is

$$m_1(x) = \frac{\lambda_2 P_0}{\lambda_1 - \lambda_2} + \frac{\Sigma_R \lambda_1 f}{(1+\alpha_{10})(\lambda_1 - \lambda_2)} \int_{x'} dx' \, m_1(x') \, K(x' \rightarrow x) \quad ;$$

where $K(x' x)$ can be shown to be $\frac{1}{2}E_1(\Sigma_T Z)$ or

$$\frac{1}{2}\int_{\Sigma_T Z}\frac{dt}{t}\mathcal{C}^{-t} \cdot$$

By replacing the equation for $M_1(x)$ with a linear system of equations, computer solutions for $M_1(x)$ were determined. Utilizing the solutions $M_1(x_i)$ for various slab thicknesses, Kr x-ray, 9.3 KeV resonant and non-resonant gamma ray currents emanating from this source surface were calculated utilizing formulations derived by Price et al. (20). Results of the calculated currents are presented as ratios of Kr K x-ray current to total 9.3 KeV gamma ray current as a function of source thickness, Figure 4.

Transport Description

The most complete manner of specifying a photon beam is to describe how many photons are going in a particular direction with a particular energy at a particular point in space. Formally, this information can be given by a flux density function of position, energy, and direction, $\Phi(r, E, \Omega)$, call it angular number flux. This function is defined so that $\Phi(r, E, \Omega)dE d\Omega$ gives the number of photons at r, with energy E about dE, going in the direction specified by the unit vector, Ω, within the element of solid angle $d\Omega$, which cross in unit time a unit differential element of area whose normal is in the direction Ω.

Before going to the equations that describe $\Phi(r, E, \Omega)$, consider the two transport problems of interest to the self-absorption problem. Assume for discussion that all recoilless photons have energy E_r, let

σ_r = absorption cross section for gammas of energy E_r

$f\sigma_r$ = recoilless absorption cross section

$\frac{f^2 \sigma_r}{1+\alpha_{10}}$ = recoilless emission after recoilless absorption cross section

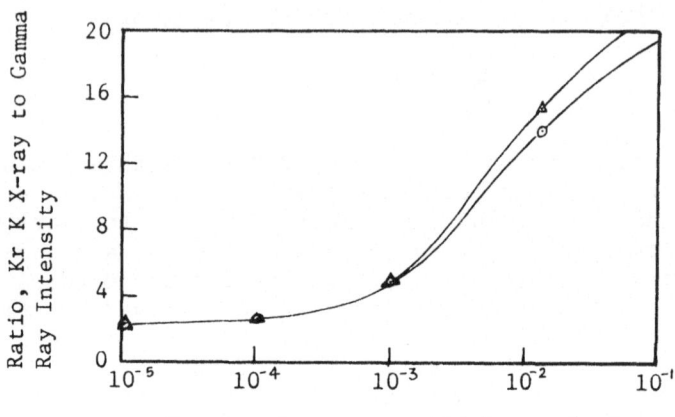

Figure 4. Theoretical ratios of Kr K x-ray to 9.3 KeV Gamma ray intensities as a function of Kr ice thickness for recoilless fraction f = 0.85.

For the case of recoilless photons only, consider a layer of thickness t of source material with S_{21} decays per cm³ per second and M_o atoms per cm³ in ground state. Consider only photons emitted in the recoilless transition 1→0. From S_{21} there results $fS_{21}/(1+\alpha_{10})$ photons emitted per cm³ per second with E_r. The emergent flux of these resonant photons from the layer can be obtained directly from standard transport theory results. The photon flux in the layer $o \leq x \leq t$ can also be obtained as $\phi_r(x)$.

Consider now the case of the recoil photons. Recoil photons arise from the daughter decays from the S_{21} decays at the rate

$$S_{21}\frac{1-f}{1+\alpha_{10}}$$

Recoil photons arise from the absorption of the E_r photons at the rate

$$f\, m_o\, \sigma_r\, \phi(\alpha)\, \frac{1-f}{1+\alpha_{10}} \ .$$

Therefore the total source of recoil photons is given by

$$S_T = \left(\frac{1-f}{1+\alpha_{10}}\right)\left(S_{21} + f\, m_o\, \sigma_r\, \phi(\alpha)\right) \ .$$

This source can also be used for a transport calculation of the emergent gamma flux.

To determine standard transport theory results, consider the standard time independent Boltzmann transport equation (21):

$$\underline{\Omega} \cdot \nabla \phi(r, E, \underline{\Omega}) + \Sigma_T(r, E)\, \phi(r, E, \underline{\Omega}) = S(r, E, \underline{\Omega})$$

$$+ \int dE' \int d\Omega'\, \Sigma(r, E' E, \underline{\Omega}'\ \underline{\Omega})\, \phi(r', E, \underline{\Omega})$$

where

$$\Sigma_T(r, E) = \Sigma_a(r, E) + \Sigma_s(r, E) = \text{the total gamma ray cross section}$$

$$\Sigma(r, E' E, \underline{\Omega}'\ \underline{\Omega}) = \text{the transfer cross section}$$

$$S(r, E, \underline{\Omega}) \quad = \text{the source term (fixed source)}$$

$$\phi(r, E, \underline{\Omega}) \quad = \text{the vector flux}$$

The ANISN Computer code (22) solves the equation for the case of interest here, i.e., one dimensional Boltzmann transport equation with isotropic, resonant (elastic) scattering for a slab by employing the diamond difference solution technique. Fourteen energy group source and cross section constants and thirty space intervals were utilized in the ANISN code calculations. ANISN results depicting ratios of Kr K x-ray to Kr γ-ray intensities as a function of Kr ice thickness are compared to ratios previously determined from consideration of Kr[83] excited state distributions, Figure 4. The diverging deviation between the two curves, (Figure 4) is attributed to some of the less rigorous approximation methods used in the excited-state formulation. Because the transport-equation formulation of self-absorption physics is the more rigorous of the two models, transport-equation results will be used later to compare with experimental results.

EXPERIMENTAL

Experimental Apparatus

Krypton ice self-absorption data are obtained by utilizing the following experimental hardware and electronic
equipment, Figure 5.

• liquid nitrogen - liquid helium Dewar

• appropriate gas handling system

• gas-irradiation vial

• self-absorption cell and associated feed line

• vacuum-pumping equipment

• proportional counter with associated electronics

The Dewar system was fashioned from a SELECT-A-STAT
Sulfrian Cryogenics, Inc., research Dewar. Basically the
Dewar consists of 2 liter helium reservoir, a 4 liter nitrogen tank providing a heat shield, and an outer vacuum con-
tainer. Additions/modifications to the Dewar included the
attachment of a helium-well bottom fitting containing a
10-mil thick beryllium window, a nitrogen-temperature shield
tailored to pass soft photons, and a room-temperature tail
section also containing a 10-mil beryllium window, Figure 6.
Emanations from the radioactive krypton disk submerged in
the lower region of the helium well, thus experience an easy
escape path to an outside positioned detector by penetrating
a thin layer of liquid helium and two 10-mil thick beryllium
windows. Because the source-absorber cell Be window is in
contact with the liquid coolant, krypton ices are always
known to be at the temperature of the liquid.

To provide sufficient gas-handling plumbing, pyrex
glass tubing, copper bellows, and stainless steel tubing
components were used throughout the system. A special
effort was made to exclude high vapor-pressure materials
(e.g., rubber hose) from the gas-handling system as any
significant degassing during vacuum tight conditions would
contaminate the otherwise high purity krypton. By operating
eight vacuum valves connecting eight vials of known volume

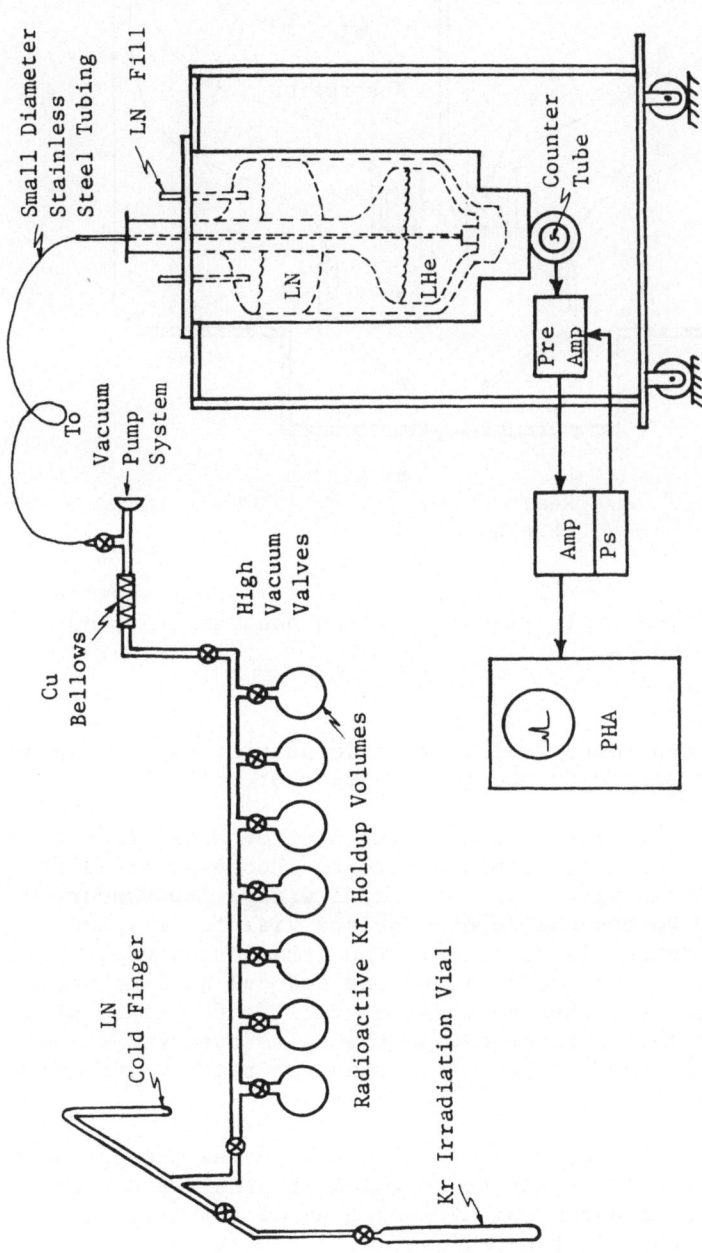

Figure 5. Complete experimental apparatus including irradiated krypton vial, pyrex-metal gas handling system krypton self-absorption cell and cryostat. Associated vacuum-pump components are not shown.

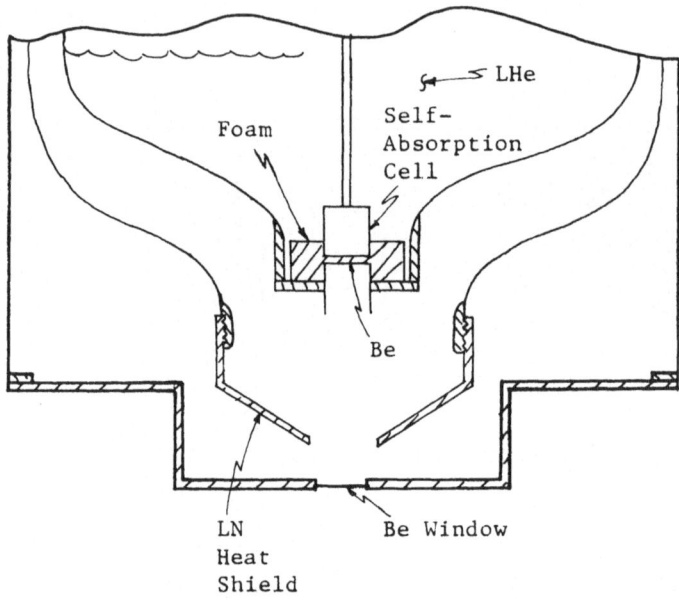

Figure 6. Lower portion of liquid helium dewar showing
self-absorption cell, liquid nitrogen heat shield, and
beryllium windows.

to the source-absorber cell, specific quantities of krypton
gas could be routed to the source-absorber cell, Figure 5.

In consideration of minimizing neutron activation in
containment combined with a desire to choose an irradiation vial fitted with a positive seal with a low vapour
pressure led to the choice of a quartz vial for krypton
containment during irradiation. The irradiation vials were
fashioned with a breakoff tip at one end and a flame remov-
able valve on the other end. After filling the vials with
known quantities of krypton gas, the valve stem was flamed
off leaving a flame sealed vial ready for reactor irradiation.

The self-absorption cell, Figure 7 serves two purposes:
it provides a cold substrate on which krypton gas solidifies
and it serves as a convenient source holder by providing a
vacuum-tight box for the radioactive ice with an easy path
for soft photons escaping towards the detector. A 10-mil

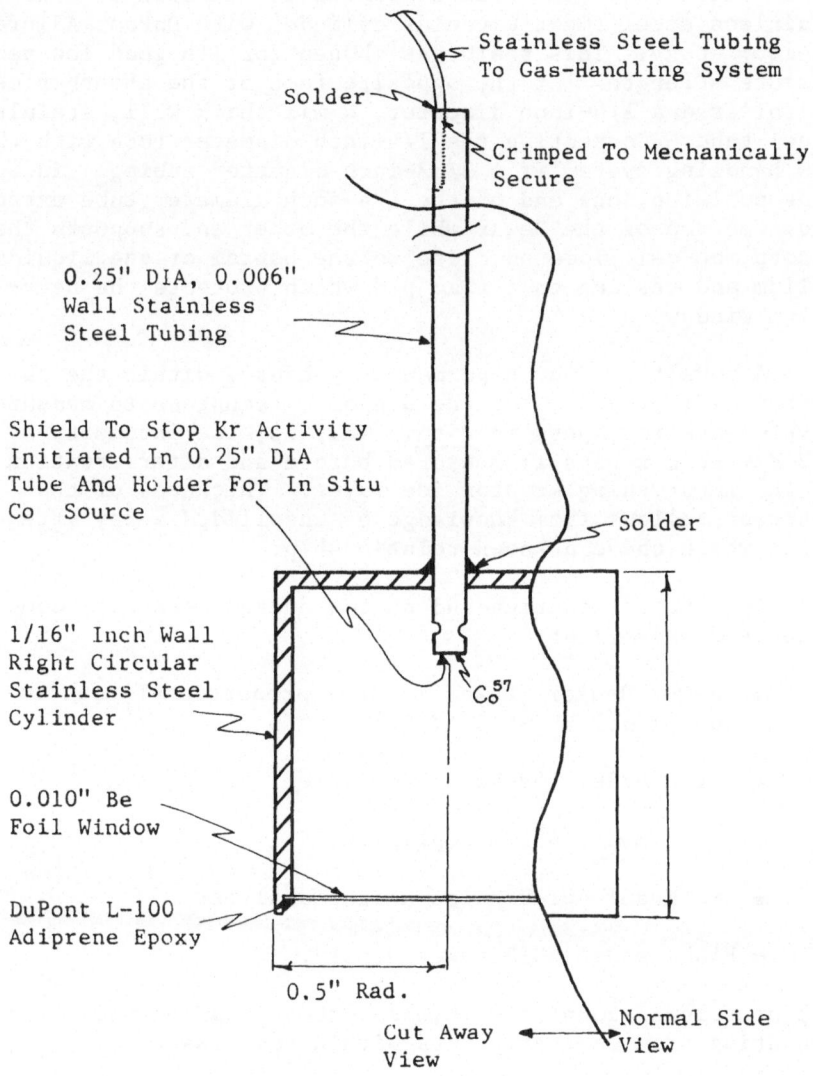

Figure 7. Self-absorption cell consisting of a hollow stainless steel right circular cylinder fitted at one end with a
10-Mil thick beryllium window and at the other end a long
1/4-inch diameter, 6-Mil wall stainless steel tubing.

beryllium foil forms the face of the absorption cell next
to the detector. The foil is bonded to the open of a hollow
stainless steel right circular cylinder with DuPont Adiprene
(L-100) epoxy. This resin was chosen for its good low temperature strength. At the opposite face of the absorber cell
is soldered a 1/4-inch diameter, 6 mil thick wall, stainless
steel tube. Connecting the 1/4-inch diameter tube with the
gas handling system is a 3/32-inch diameter tubing. In opera
ting position, one end of the 1/4-inch diameter tube extends
from the top of the Dewar while the other end supports the
absorption cell located close to the bottom of the liquid
helium and resting on a foam pad which protects the beryllium window.

A cobalt-57 source permanently housed within the absorber cell provides for the use of a technique to measure
krypton ice thickness in situ. That is, the intensity of
6.3 KeV Fe K x-rays is measured before and after creation
of the intervening krypton ice layer. Thickness of the
absorber follows from knowledge of uncollided x-ray attenuation which obeys a known relationship.

The photon counting and photon energy selecting components consisted of:

- Xe-CH$_4$ Reuter-Stokes (1 atm) proportional counter
  tube model RSG-61S (J749)

- Ortec Model 109-PC preamplifier

- Ortec Model 435-A amplifier

- TMC Model 401-D pulse height analyzer

- Fluke model 405B power supply.

A proportional-counter shield absorbed unwanted radiations
emanating from scatter points within the Dewar.

Computer Analysis of Data

Experimental data (PHA 400 channel data) was scruti-
nized for Fe 6.4 KeV x-radiation, Kr 9.3 KeV gamma, and

Kr 12.6 KeV x-radiation. The computer analysis of the data consisted of generating a rational function with coefficients assigned to give the "best" approximation to the set of discrete data in the sense of minimizing the least squares of differences between the data points and the candidate function.

The least squares routine is utilized by first determining the rational function "best" representing selected discrete data points in the largest Kr K x-ray peak. The larger intensities in this peak combined with its highest energy position of the three-peak set, make it the best candidate for containing data points unaffected by wing data points of a neighboring peak.

After the closed form of the rational function is determined, the Kr K x-ray peak is reconstructed utilizing the rational function and DO looping the set of independent-variable values representing the full wing-spread of the peak. This set of artificial discrete points is then subtracted from the complete 400 data points contained in the experimental spectra. With this stripping operation complete, a new set of artificial points is determined for the peak next lower in energy, i.e., the Kr γ-ray peak. Now stripped of influences from the larger Kr K x-ray peak, the Kr γ-ray peak is fitted with a function most representative of the intensities. Again, artificial discrete points are generated representing full wing spread of the Kr γ-ray peak and then subtracted from the remaining experimental data set already voided of Kr K x-ray peak data.

The above procedure is performed a third time to determine the curve-fit parameters and artificial discrete data points representing the Fe K x-ray peak (recall that this peak intensity is used for source thickness measurements). If not sufficient Fe K x-ray intensities are present in the PHA spectra, the rational function thus generated deviates significantly from a Gaussian-like shape, e.g., becomes negative in one peak wing and monotonically increases in the opposite peak wing. Fits of this nature were rejected. Peak intensities are obtained by merely summing all discrete points generated artifically by use of the computer determined rational function.

RESULTS

All original data were obtained in 27 runs, 12 of which represented different krypton ice thicknesses. For quali- tative analysis here, the results of three counting sequence: are depicted by plotting count rate as a function of channel number with intensities normalized to peak Kr K x-ray intensities, Figure 8.

Referring to Figure 8, consider the Kr γ-ray peak heights and the calculated source thickness shown in each plot. It is seen for the greater source thicknesses that the size of the Kr γ-ray peak decreases. This result is expected as predicted by previously presented mathematical analysis.

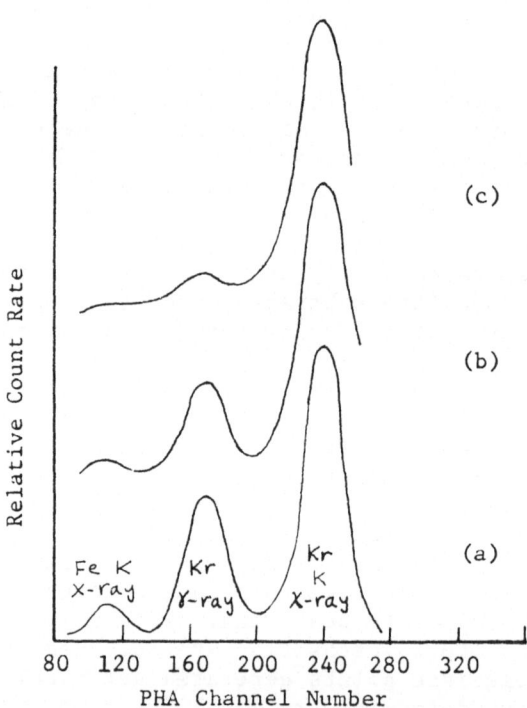

Figure 8. Original PHA data, (a) source in gaseous state, (b) Kr ice thickness = 0.0022 cm, (c) Kr ice thickness = 0.016 cm.

   In the quantitative computer analysis of the PHA
spectra, the following items were determined.

- Kr K x-ray intensity

- Kr 9.3 KeV γ-ray intensity

- Ratio Kr K x-ray to Kr γ-ray intensities

- Kr ice source thickness as determined from Fe x-ray
  attenuation data and Kr x-ray intensity data

- Kr K x-ray intensity data

Kr K x-ray and Kr γ-ray intensities were corrected for
absorption through Dewar parts and proportional counter
efficiency. Experimental data, i.e., results of computer
analyzed PHA data in the representation of ratios of Kr
x-ray to Kr 9.3 KeV γ-ray intensities, are plotted as
points, Figure 9.

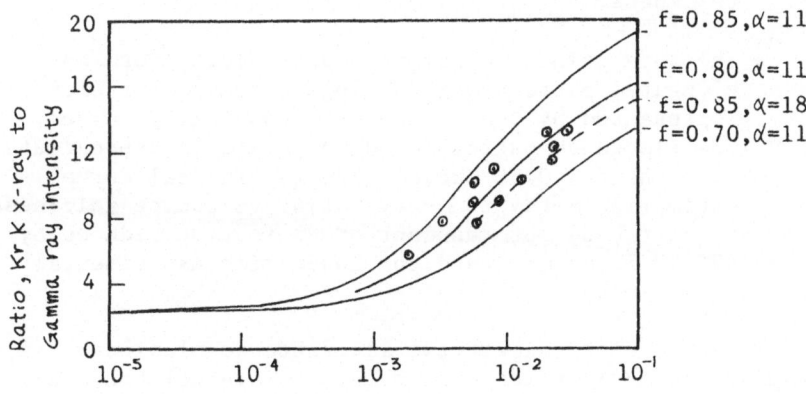

Figure 9. Ratios of Kr x-ray to Kr γ-ray intensity from
transport equation predictions (solid and dashed lines)
with experimental data points superimposed.

Curves depicting x-ray to γ-ray ratios as a function of Kr source thickness as calculated from transport-equation predictions are superimposed to depict their relationship with experimental results. Referring to Figure 9, the following comments can be made:

- All experimental data points are representative of Kr ice thickness ranging from about 1.5×10^{-3} cm to 2.2×10^{-2} cm thickness. Although this does not represent a wide range of source thicknesses, experiments with thinner or thicker sources pose problems. Thinner sources, although experimentally feasible, yield results that are less significant because dependence of the Kr x-ray to γ-ray ratio is less apparent. Thicker sources, on the other hand, are difficult to produce experimentally. The relatively poor heat conductivity of Kr ice combined with heat of deposition introduced to the Kr ice at the ice-gas interface necessitates an ever increasing freezing time for thicker ices. When longer ice formation times are used it is likely that undesirable Kr ice will form on lower conductivity sections of the absorption cell, i.e., the stainless steel surfaces. This maverick source ice introduces errors in the data because of its unknown position and thickness.

- The data points, considered as a block, form an elongated pattern with a slope close to the slope represented by the f = 0.7 to 0.85 family curves. The elongated pattern of data points is approximately over the f = 0.80 curve. The theoretical curve estimating x-ray to γ-ray ratios <u>vs</u> source thickness for f = 0.85 but resonant cross-section reduced by 40% is located toward the lower side experimental data block.

These results suggest two alternate conclusions: (a) assuming the earlier measured value of internal conversion coefficient is correct (i.e., α = 11) the results of this study suggest that the Mössbauer recoilless fraction of solid krypton at LHe temperature is 0.80 $\pm$ 0.05. (b) assuming the recently measured value of the internal conversion coefficient is correct, i.e., α = 18, the results of this study suggest that an f value of a least 0.85 for solid krypton at LHe temperature. Uncertainty values attached to the f value for alternate (a) above (i.e., $\pm$

0.05) are suggested by the certainment of all experimental
data points by the f = 0.85 and f = 0.75 family curves.
Because for alternate conclusion (b) above, the theoretical
curve for f = 0.85 and α = 18, is located toward the lower
region of the data pattern, the recoilless fraction f,
assuming α = 18, suggested by the data is slightly greater
than 0.85.

For a source disk diameter of 2.54 cm and an ice
density of 3.095 gm/cm^3 the thickest source (0.021 cm)
corresponds to a collection of about 2.38 x 10^{21} atoms of
total krypton. Comparing this to the 3.65 x 10^{21} atoms
that were introduced into the irradiation vial (calculated
from known cc-atm. of gas in the irradiation vial) this
corresponds to a collection efficiency of 65%. In their
paper, Pasternak and Sonnino (7) reported a maximum deposition of 80%. Although experimental apparatus were
different, the 65% collection efficiency experienced in
this experiment seems low. The comparatively low collection efficiency experienced in this experiment is likely
due to short deposition times utilized. However, a deli-
berate extended deposition time for the collection of Kr
crystals on the beryllium window, would have increased the
probability of depositing unwanted ice on the absorber cell
stainless steel side and top surfaces.

The almost unpreventable deposition of some Kr
crystals in the absorber-cell inside surfaces (other than
beryllium window) probably accounts for some of the spread
in data points assigned to the same or near same ice thicknesses. That is, Kr x-ray intensities emanating from radioactive ice deposited on absorber-cell innards other than
the beryllium window will result in inaccurate determinations of ice thickness assumed existing on the beryllium-
window-substrate only. In addition, unknown thicknesses
of maverick ice result in an alteration of the Kr x-ray to
γ-ray ratio that is incalculable with practical considerations.

It is interesting to note that the Kr γ-ray intensities
detected via the self-absorption experiment are akin to the
γ-ray intensities detected in a conventional Mössbauer
experiment when the relative doppler shift is zero. A
difference does exist in the nature of the source, however.
In the self-absorption experiment the source parent atoms
(Kr^{83m}) are uniformly distributed throughout the absorber;

in the conventional experiment, however, source and absorber
atoms are uniformly distributed throughout one half space
and only absorber and scattering atoms exist in the second
half space.

Consider the sensitivities of the two experiments
(self absorption and conventional Mössbauer) to the parameters recoilless fraction, f, and internal conversion, α.
As f approaches zero, emanations of Kr γ-rays are at a
maxiumum in the self-absorption experiment. That is, all
γ-rays experience only electronic collisions when f = 0.
In the conventional Mössbauer experiment as f approaches
zero, off-resonance (large Doppler velocity) and on-
resonance transmission levels become equal. Assuming that
the source thickness is not thin (i.e., $T_s \approx 1$) but is
accurately known in both experiments, the self-absorption
experiment and the conventional Mössbauer experiment should
yield results equally sensitive to f values. This equal
sensitivity to f value is due to the fact that both
experiments utilize "on-resonance" data (emanation intensity
from the solid source in the self-absorption experiment and
minimum transmitted intensity in the conventional experiment
and "off-resonance" data (emanation intensity from the
gaseous source in the self-absorption experiment and maximum transmitted intensity in the conventional experiment).
If source and absorber are both thin, however, i.e., $T_s \ll 1$
and $T_A \ll 1$, then the f value can be determined from conventional Mössbauer data. As indicated earlier, for the
case of thin sources, self-absorption data become insensitive to changes in f and are thus less effective in determining f than conventional Mössbauer data.

The internal conversion coefficient, α, like the
recoilless fraction, f, enters into the calculation of the
resonant cross-sections. Changes in α however do not
affect the intensity of escaping γ-rays from a Mössbauer
source to the degree that changes in f affect intensities.
This is true because changes in α affect only the total
gamma ray source strength and resonant cross section values.
Whereas changes in f affect non-resonant and resonant source
strengths and resonant cross sections. Since cross sections
for non-resonant and resonant interactions differ greatly,
changes in f result in greater changes to emergent total
gamma ray flux from a thick Mossbauer source than do equal
percentage changes in α .

The reasons for the apparent greater sensitivity of the conventional Mössbauer experiment to changes in the internal conversion coefficient α, Figure 1, than the self-absorption Mössbauer experiment to changes in α, Figure 9, are not immediately obvious. Changes in f according to Kolk (1) approach 45% for a 64% change in α. According to this work, for a 64% change in α, f is changed by approximately 15%, Figure 9.

ACKNOWLEDGEMENTS

Equipment and partial support for this work have been supplied by The Ohio State University, Department of Nuclear Engineering and Battelle Columbus Laboratories. The author is also grateful to Dr. Robert A. Krakowski of Los Alamos Scientific Labs and Dr. Robert F. Redmond of the The Ohio State University for their many hours of valuable discussions.

REFERENCES

1. Kolk, B., "Mössbauer Recoilless Fraction of Kr^{83} in Solid Krypton", Phys. Letters 35A:83-84 (1971).

2. Kolk, B. et al., "Internal Conversion Measurements on Kr^{83}," Nucl. Phys. A(Netherlands), A194: 614-624 (1972).

3. Ruby, S. L., et al., "Lifetimes of the Low-Energy M1 Transitions in La^{137} and Kr^{83}", Phys. Rev. 29:826-828 (1963).

4. Hazony, Y., et al., "Mössbauer Effect in Kr^{83} Trapped in a Clathrate Compound", Phys. Letters 2:337-339 (1962).

5. Ruby, S. L. and Selig, H., "Mössbauer Study of Kr^{83} in the Compound KrF_2", Phys. Rev. 147:348-354 (1966).

6. Pasternak, M., et al., "The Mössbauer Effect in Solid Krypton", Phys. Letters 22:52-53 (1966).

7. Pasternak, M. and Sonnino, T., "Mössbauer Effect Studies on Kr^{83} in Bromide and Bromate Crystals", Phys. Rev. 164:384-390 (1967).

8. Bukshpan, S., et al., "Mössbauer Effect in Kr^{83} with
    Sources of Se^{83}", Phys. Letters 27A:372-373 (1968).

9. Gilbert, K. G., "Mössbauer Effect in Solid Krypton",
    UCRL - 50474, Lawrence Radiation Laboratory,
    University of California, Livermore, TID - 4500,
    UC - 34 (1968).

10. Gilbert, K. and Violet, C. E., "Mössbauer Recoilless
    Fraction of Solid Krypton:, Phys. Letters 28A:285-
    286 (1968).

11. Brown, J. S. and Horton, G. K., "Model Potentials and
    the Dispersion Law in Solid Krypton", Phys. Rev.
    Letters 18:647-649.

12. Private Communications between Gilbert, K. and Violet,
    C. E. with Jelinek, G., Reference 9.

13. Brown, J. S., "Anharmonic Effects on the Mössbauer
    Recoilless Fraction of Solid Krypton", Phys. Rev.
    187: 401-402 (1969).

14. Brown, J. S., "Mössbauer Recoilless Fraction of Solid
    Krypton II", Phys. Rev. B 3 21-24 (1970).

15. Mahesh, K., "Debye Model Calculations of the Mössbauer
    Effect of 9.3 KeV Transition of Kr^{83} Solid", J.
    Physical Soc. Japan 28:818-822 (1970).

16. Windecker, R. C., Thesis 1970, Illionis University,
    Urbana (COO-1198-762).

17. Frauenfelder, H., The Mossbauer Effect, W. A. Benjamin,
    Inc., New York (1963).

18. Wertheim, G. K., Mössbauer Effect: Principles and
    Applications, Academic Press, New York (1964).

19. Margulies, S. and Ehrman, J. R., "Transmission and
    Line Broadening of Resonance Radiation Incident on
    a Resonance Absorber", Nuclear Instruments and
    Methods 12:131-137 (1961).

20. Price, B. T., Radiation Shielding, Pergamon Press,
    New York (1957).

21. Meghreblian, R. V. and Holmes, D. K., Reactor Analysis,
     McGraw-Hill Book Company, Inc., New York (1960).

22. Soltesz, R. G., "ANISN, A One Dimensional Discrete
     Ordinates Transport Code:, RSIC Computer Code
     Collection", Contributed by Westinghouse Astronuclear
     Laboratory, Pittsburgh, Penn., WANL - TMI - 1967,
     Oak Ridge National Laboratory, CCC-82-F (April 1969).

MÖSSBAUER SPECTROSCOPY OF ^{125}Te - SOME NEW RESULTS AND APPLICATIONS

P. Boolchand,[*] B. B. Triplett and S. S. Hanna

Department of Physics
Stanford University
Stanford, California 94305[†]

and

J. P. deNeufville

Energy Conversion Devices, Inc.
Troy, Michigan 48084

Narrow and reproducible linewidths for the 35.5-keV Mössbauer resonance in ^{125}Te have been observed with a 2.7-year ^{125}Sb source diffused in Cu. Measurements made on cubic ZnTe as a function of absorber thickness reveal that the linewidth of these sources is $\Gamma_{source} = 5.20 \pm 0.08$ mm/sec. This result is in good agreement with the minimum observable linewidth $2\Gamma_{nat}$ as calculated from the measured half-life of the 3/2 level, $t_{1/2} = 1.475 \pm 0.010$ nsec. From this Mössbauer measurement the Debye temperature of cubic ZnTe is found to be 175 ± 9 °K. The Mössbauer effect in ^{125}Te provides a useful technique for characterizing amorphous chalcogenide semiconductors. Investigations on sputtered thin films of amorphous Ge_xTe_{1-x} were performed as a function of composition, temperature, and heat treatment. The variation of the quadrupole splitting as a function of composition exhibits discontinuities in slope at $x = 0.33$ and

* Permanent address: University of Cincinnati, Cincinnati, Ohio 45221
† Supported in part by the National Science Foundation.

0.50. These results are indicative of chemical ordering in the amorphous Ge_xTe_{1-x} system at these compositions. A structural model of amorphous Ge_xTe_{1-x} alloys based on two inequivalent Te sites is developed for the Te rich phase and is found to fit the data well. Annealing of the films is shown to lead to a greater degree of structural order on a microscopic scale. Crystallizing the films is shown to lead to phase separation of the system into crystalline GeTe and Te metal.

INTRODUCTION

Investigations of the 35.5-keV gamma resonance in ^{125}Te have primarily been based on the use of ^{125}I ($t_{1/2}$ = 60 days) in Cu and ^{125m}Te ($t_{1/2}$ = 58 days) in ZnTe as sources of monoenergetic gamma rays. Both these sources are known to exhibit broad linewidths, and in particular the linewidths of ^{125}I sources are not reproducible. We now report that recent experiments [1,2] with ^{125}Sb (2.7 years) sources diffused in Cu demonstrate that these sources are characterized by a narrow and reproducible linewidth. Measurements [2] made on cubic ZnTe as a function of absorber thickness reveal that the linewidth of these sources agrees well with $2\Gamma_n$, the minimum observable linewidth based on the half life of the 3/2 level measured directly from its rate of decay [3]. These experiments have also shown that the Mössbauer f factor of ^{125}Sb sources is substantially larger than for ^{125}I sources. The combination of these favorable properties, viz., narrow linewidth, long parent half-life, and large f factor, makes an ^{125}Sb source particularly attractive for future Mössbauer effect work with ^{125}Te. The isomer shift of Te metal relative to this source is found to be +0.50 ± 0.04 mm/sec. We propose this source be used as a standard for ^{125}Te isomer shift measurements.

The significance of a narrow single-line source cannot be overemphasized for measurements of electric and magnetic hyperfine interactions. The ^{125}Te resonance has a natural linewidth $2\Gamma_n$ of 5.20 ± 0.04 mm/sec and the presence of hyperfine interactions lead invariably to partially resolved spectra. Reliable analysis of such spectra is possible only by giving due consideration to the linewidths of the unresolved components [4]. In this sense the present results offer the promise of higher precision in the study of the ^{125}Te gamma resonance.

A particularly important application of ^{125}Te spectroscopy is its use as a microscopic probe in the characterization of chalcogenide-amorphous semiconductors. The applied interest in these non-crystalline materials derives from the Ovonic Threshold Switching that these substances display. The nature of the bonding and structure in the amorphous Ge_xTe_{1-x} system has for this reason led to numerous investigations [5] involving a variety of techniques such as X-ray [6], neutron RDF [7], and XPS [8]. In this paper we present results on this system obtained with ^{125}Te Mössbauer spectroscopy. Amorphous thin films of Ge_xTe_{1-x} were used as absorbers, and experiments were performed as a function of composition and temperature of the films. Spectra of virgin and annealed films were also obtained to investigate the role of thermally induced changes in the structure of amorphous alloys. The amorphous films were crystallized by heating in vacuum above the crystallization temperature T_x. Crystallization of these films was found to produce crystalline GeTe and Te from the amorphous Ge_xTe_{1-x} system. The experimental procedure and results of the present investigations are summarized in the next section, where we also develop a structural model of amorphous Ge_xTe_{1-x} alloys in order to interpret the data. Substantial changes in the Mössbauer effect spectra of virgin and annealed amorphous Ge_xTe_{1-x} films suggest the existence of defect structures in the virgin films. It is believed that the process of annealing leads to formation of a greater degree of structural ordering.

The structure of amorphous Ge_xTe_{1-x} alloys has been the subject of numerous other physical measurements. Also, in the next section an attempt is made to discuss the existing semiclassical description of these alloys with the model developed from the Mössbauer measurements.

Predictions for the amorphous phase of Te, corresponding to composition $x = 0$ in Ge_xTe_{1-x}, are of particular interest. For amorphous Te, one infers a substantially larger quadrupole splitting and a smaller isomer shift than the value known for the crystalline form.

PROCEDURE, RESULTS, AND INTERPRETATION

Mössbauer effect spectra were accumulated by use of a constant acceleration drive with provisions for a simultaneous velocity calibration. The ^{125}Te spectra were recorded with an ^{125}Sb(Cu) source by detecting the K_α escape peak (5.8 keV) in a $\overline{Xe}$ filled proportional counter.

Linewidth Measurements on a ^{125}Sb(Cu) Source

Mössbauer spectra from an ^{125}Sb(Cu) source with ZnTe absorbers of different thicknesses were recorded. By linear extrapolation to zero absorber thickness one can eliminate contributions to the observed linewidth from absorber thickness. The cryostat used in these measurements provided both source and absorber cooling down to liquid nitrogen or helium temperatures by means of exchange gas system. Samples of ZnTe obtained from three suppliers were investigated separately. Sample A consisted of chips of cubic single crystals from Gould Industries, Cleveland, Ohio; sample B consisted of polycrystalline ZnTe, stated purity 99.999%, bought from Research Inorganic Chemical Corporation, Sun Valley, California; and sample C consisted or enriched $Zn^{125}Te$ hot pressed in a lucite matrix and was made available by New England Nuclear, Boston, Massachusetts.

The narrowest linewidths were observed for ZnTe absorbers made from sample A. These experiments were performed at 78°K and 4.2°K on the same set of absorbers. A summary of the linewidth data obtained as a function of absorber thickness appears in Fig. 1. The data were fit to a straight line and gave the following results.

T = 4.2°K Γ_{source} = 5.30±0.05 mm/sec f_{ZnTe} = 0.57±0.01

T = 78°K Γ_{source} = 5.10±0.08 mm/sec f_{ZnTe} = 0.32±0.02

The values of Γ_{source} are in good agreement with $2\Gamma_n = \hbar/\bar{\tau} = 5.20 \pm 0.04$ mm/sec obtained from the measured [3] half life $t_{1/2} = 1.475 \pm 0.010$ nsec of the $3/2^+$ state. In a Debye model the measured f values of cubic ZnTe translate to a mean $\theta_D = 175 \pm 9$°K. Recently Blattner et al. [9] have reported an effective Debye temperature of $\theta_M = 180 \pm 6$°K for cubic ZnTe from X-ray measurements.

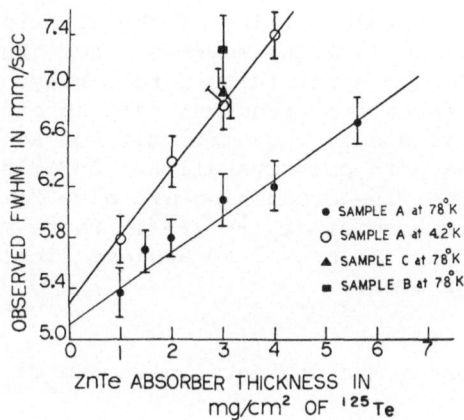

Fig. 1. Observed linewidth of a ^{125}Sb(Cu) source as a function of ZnTe absorber thickness in mg/cm^2 of ^{125}Te.

Sizable line broadening was observed in polycrystalline ZnTe samples B and C. For an absorber of ^{125}Te 3 mg/cm^2 thick, the linewidths for samples B and C were found to be 20% broader than those for sample A (Fig. 1). An X-ray examination of samples A and B showed additional peaks in sample B which can be identified as the hexagonal phase of ZnTe. The line broadening in sample B may thus be attributed to the existence of an unresolved quadrupole splitting in a non-cubic phase. Since the color of the cubic sample A was dull orange in contrast to the brown color or the other two samples, it is likely that the above explanation also applies to sample C.

The narrower emission linewidths of ^{125}Sb(Cu) sources compared with those of ^{125}I(Cu) and Zn^{125m}Te sources also deserve a few comments. It is known that Sb dissolves in Cu, although the same cannot be said for I in Cu. It is likely that CuI is formed during the electroplating and diffusion processing of I in Cu sources. It is known that CuI exists both in a cubic and non-cubic phase [10]. These observations suggest a possible explanation

for the lack of reproducibility of the linewidths that have
been observed with $^{125}I(\underline{Cu})$ sources. For $Zn^{125m}Te$
sources on the other hand, failure to achieve natural linewidths, in the absence of resonant self-absorption in the
source matrix, is due to the difficulty of achieving a
purely cubic phase in polycrystalline ZnTe samples. These
considerations on line broadening are also relevant to the
Mössbauer spectroscopies of $^{127,129}I$ where matrices of
ZnTe and Cu have been used to achieve single line sources
or absorbers, or both.

Mössbauer Effect Measurements on the Ge_xTe_{1-x} System

Thin amorphous films of Ge_xTe_{1-x} were used as absorbers
in conjunction with $^{125}Sb(\underline{Cu})$ sources. The measurements
were performed under a variety of conditions which may be
characterized by the following variables:

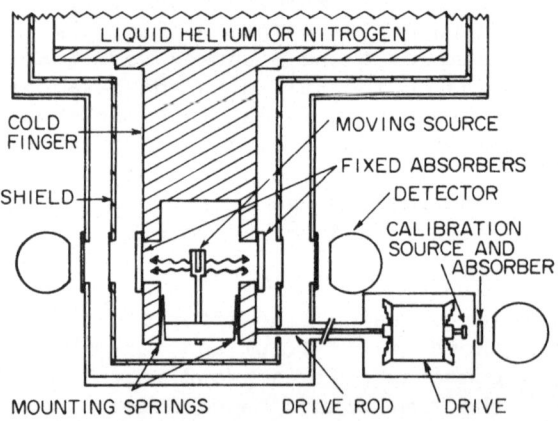

Fig. 2. Schematic drawing of the cold-finger cryostat and
detector arrangement.

    (i)   Temperature

   (ii)   Composition

  (iii)   Heat treatment: annealing effects on amorphous films, and amorphous to crystalline phase transformations.

    A novel feature of the method used in these measurements provided for calibrating the drive while simultaneously recording spectra of two absorbers maintained at cryogenic temperatures. Figure 2 illustrates the schematic arrangement of the experiment. The two absorbers and source were cooled by conduction by means of a Cu cold finger. The source was driven horizontally with a loudspeaker drive assembly which was calibrated at the far end by recording spectra from a ^{57}Co($\underline{Pt}$) source and 1 mil Fe metal absorber. Figure 3 reproduces spectra of K_2TeO_3 and Te metal obtained in a single run with the simultaneous Fe calibration. The present method is particularly attractive for comparing isomer shifts of different Te absorbers. Parameters derived from these data are summarized in Table I.

Table I. Parameters Derived from the

Data in Figure 3

Absorber	Thickness mg/cm^2	Γ_{exp} mm/sec	Δ mm/sec	$\delta^\dagger$ mm/sec
ZnTe	2.80	6.50 ± 0.14	-	−0.15 ± 0.04
K_2TeO_3	4.00	6.52 ± 0.10	6.80 ± 0.05	+0.28 ± 0.05
Te	4.00	6.38 ± 0.11	7.78 ± 0.06	+0.52 ± 0.05

†The isomer shift δ is measured relative to the source, ^{125}Sb in Cu.

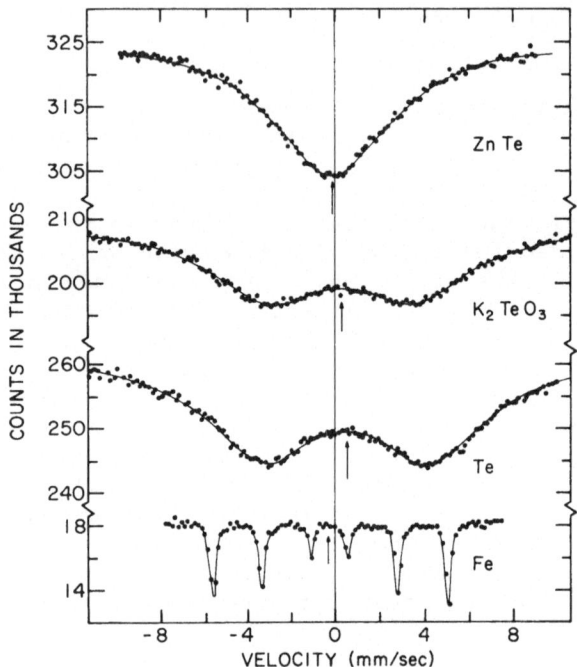

Fig. 3. Mössbauer spectra accumulated with a ^{125}Sb(<u>Cu</u>) source and the absorbers ZnTe, K$_2$TeO$_3$, and Te. Also shown at bottom is a spectrum taken with a ^{57}Co(<u>Pt</u>) source and a 1 mil Fe metal absorber.

Ge$_x$Te$_{1-x}$ Absorbers

The amorphous films used in the present investigations were prepared at Energy Conversion Devices, Troy, Michigan by rf sputtering the material onto 15 µ thick Al foil substrates. Typical film thicknesses ranged from 5 to 50 µ. For the Ge rich composition, absorbers were prepared by stacking together a few films to achieve sufficient ^{125}Te thickness.

Temperature Measurements

The bulk of the Mössbauer effect measurements were performed at liquid helium temperature. In the preliminary phase of these experiments, however, some measurements were performed at 78°K to investigate whether the choice of substrate might produce significant changes in the spectra of amorphous Ge_xTe_{1-x} films. Figure 4 reproduces spectra of films sputtered on glass and Al substrates. In each case the spectra of the amorphous films were characterized by a weakly split quadrupole doublet which showed no significant substrate dependence either in splitting or linewidth.

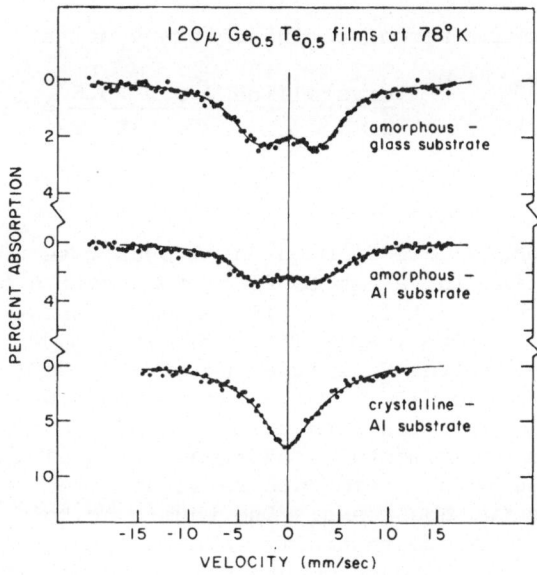

Fig. 4. Mössbauer spectra of 120 μ $Ge_{0.5}Te_{0.5}$ taken at 78°K. The upper two spectra were taken with amorphous films on glass and aluminum substrates, respectively. The lower spectrum was taken after crystallizing (see text) the film on the Al substrate.

Table II. Summary of ^{125}Te Mössbauer Effect
Results on 120 μ Thick $Ge_{0.5}Te_{0.5}$ Absorbers at 78°K

Absorber	Sub-strate	Γ_{exp} (mm/sec)	% Effect	Area[a]
Amorphous	glass	5.73 ± 0.20	1.95 ± 0.05	22.3 ± 1.0
Amorphous	Al	5.63 ± 0.20	2.28 ± 0.04	25.6 ± 1.0
Crystalline	Al	6.28 ± 0.17	7.13 ± 0.11	44.8 ± 1.5

[a] Area under the resonance in arbitrary units. Note that for the Al substrate: $\dfrac{f_{crystalline}\ Ge_{0.5}Te_{0.5}}{f_{amorphous}\ Ge_{0.5}Te_{0.5}} = 1.75 \pm 0.13.$

An amorphous GeTe film on an Al substrate was crystallized by heating in vacuum to 250°C for about 6 hrs. A
spectrum of the crystallized film (Fig. 4) showed a single
line with a 7.1% dip (Table II). The area under the resonance for crystalline GeTe is a factor of 1.75 larger than
that for amorphous GeTe. These measurements were performed
in identical geometries with absorbers of the same thickness.
For identical experimental conditions it is clear that the
ratio of areas in the resonance reflects directly the ratio
of f in the two absorbers. One thus concludes that f in
amorphous GeTe is substantially lower than that in crystalline GeTe at 78°K. Measurements performed at liquid helium
temperature show, however, that the f factors in amorphous
and crystalline GeTe are approximately equal. We believe
the strong temperature dependence of f in amorphous GeTe
may be a manifestation of the resonant atom rattling in its
site. This subject merits further study.

In this connection, recent measurements of the heat
capacities of amorphous and crystalline Ge have shown them
to be closely proportional at all temperatures below 30°K.

This result has led to the rather surprising suggestion [11]
that the phonon density of states spectrum in amorphous Ge
closely resembles the spectrum for the crystalline phase
(but with all frequencies in the amorphous phase reduced by
16% in order to account for the ratio of the heat capacities).
The strong temperature dependence of the ratio of the f
values of amorphous and crystalline GeTe allows an immediate
statement that for GeTe there cannot be such a similarity
in the density of states spectra. Mössbauer f values
probe a lower frequency region of the phonon spectrum than
do heat capacity measurements [12]. For this reason, f
values and heat capacities, together with Raman scattering
and electronic tunneling, provide powerful, complementary probes of the structure of the vibrational spectra of
amorphous materials.

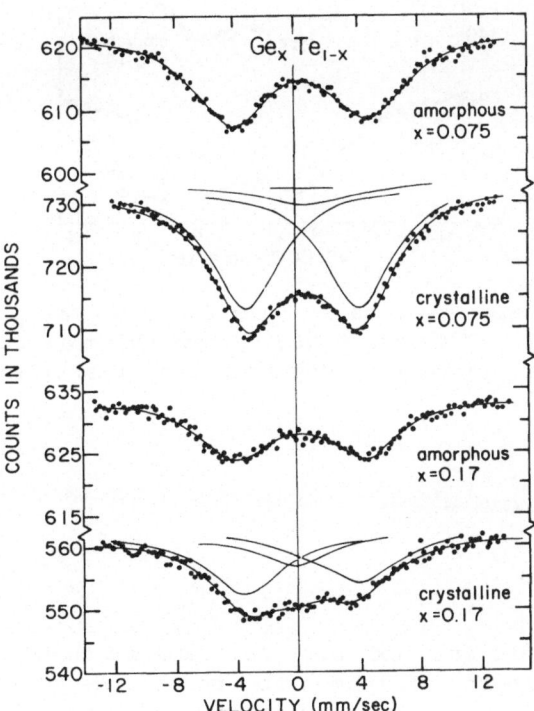

Fig. 5. Comparison of Mössbauer spectra taken on the amorphous
and crystalline phases obtained from Ge_xTe_{1-x} for x = 0.075
(upper pair of spectra) and x = 0.17 (lower pair of spectra).

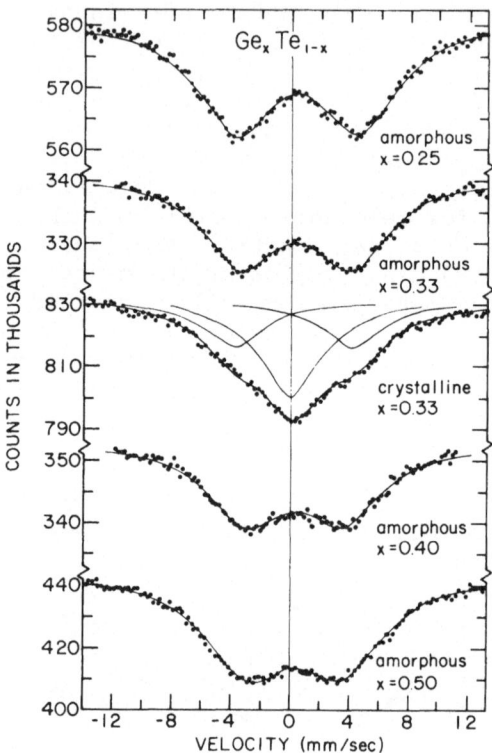

Fig. 6. Mössbauer spectra for amorphous Ge_xTe_{1-x} with x = 0.25, 0.33, 0.40 and 0.50. Shown for comparison in the middle of the figure is a spectrum for crystalline Ge_xTe_{1-x} with x = 0.33.

Our measurements exhibit the dramatic change in Te coordination in the amorphous to crystalline phase transformation of GeTe. The linewidth of the crystalline GeTe spectrum has now been investigated [13], and it cannot entirely be attributed to the effect of absorber thickness. One estimates that 6% of the linewidth in crystalline GeTe results from the presence of a non-vanishing electric field

gradient.   This result is corroborated by X-ray examination
of our sample which showed a slight (1.5%) rhombic distortion of the cubic unit cell.

   The appearance of a quadrupole splitting (QS) in amorphous GeTe shows that the local coordination at Te in this
host is less symmetric than in crystalline GeTe.   In the
amorphous GeTe spectrum, the linewidths of the components
are quite broad and we attribute this to a distribution of
QS's.   This distribution may be due to inhomogeneous

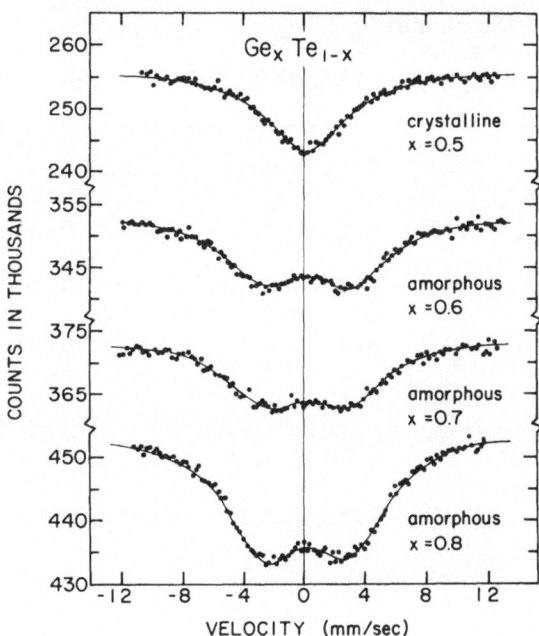

Fig. 7. Mössbauer spectra for amorphous Ge_xTe1-x with
x = 0.6, 0.7, and 0.8 compared with crystalline   GeTe
(upper spectrum).

distortions of the local coordination around the Te sites. These observations can be reconciled in terms of a three fold coordinated model for amorphous GeTe and will be discussed later on.

Composition

Spectra of freshly sputtered Ge_xTe_{1-x} films (virgin) are presented in Figs. 5, 6, and 7 sequentially with increasing Ge fraction. In all cases the spectra consisted of partially resolved doublets and were fit to a superposition

Table III. Parameters Derived from the Data in

Figs. 5, 6, and 7 for Amorphous Films

Ge_xTe_{1-x} Absorber	Thickness mg/cm^2	Γ_{exp} mm/sec	Δ mm/sec	$\delta^\dagger$ mm/sec
x = 0.075	1.45	6.14 ± 0.11	9.34 ± 0.07	+0.38 ± 0.07
x = 0.17	0.53	6.58 ± 0.20	8.84 ± 0.11	+0.42 ± 0.12
x = 0.25	3.40	6.34 ± 0.10	8.42 ± 0.06	+0.30 ± 0.06
x = 0.33	3.17	6.46 ± 0.08	7.80 ± 0.04	+0.27 ± 0.05
x = 0.40	1.94	6.49 ± 0.11	6.85 ± 0.08	+0.26 ± 0.08
x = 0.50	1.07	6.60 ± 0.06	6.17 ± 0.04	+0.22 ± 0.03
x = 0.60	1.31	6.22 ± 0.12	6.17 ± 0.08	+0.16 ± 0.08
x = 0.70	1.43	6.17 ± 0.16	5.57 ± 0.08	+0.17 ± 0.08
x = 0.80	2.8	6.23 ± 0.10	5.68 ± 0.08	+0.22 ± 0.06

$\dagger$ The isomer shift δ is measured relative to the source, ^{125}Sb in Cu.

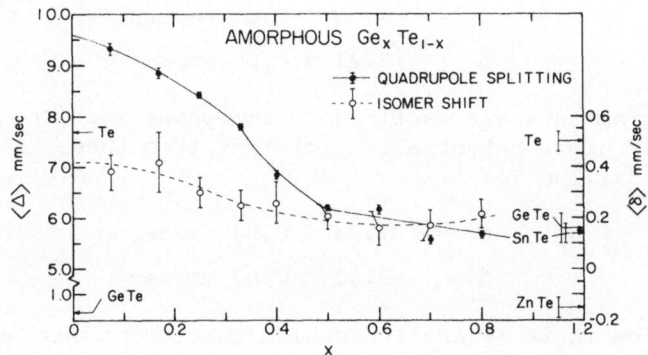

Fig. 8. Mean quadrupole splitting $\langle\Delta\rangle$ and isomer shift $\langle\delta\rangle$ for virgin amorphous Ge_xTe_{1-x} as a function of x.

of two Lorentzian line shapes to extract a mean quadrupole splitting $\langle\Delta\rangle$, isomer shift $\langle\delta\rangle$, and observed linewidth $\langle\Gamma_{exp}\rangle$ (Table III). The trends of $\langle\Delta\rangle$ and $\langle\delta\rangle$ as a function of composition are shown in Fig. 8.

The most striking features of these trends are the rapid and monotonic increase in $\langle\Delta\rangle$ for the composition range $x < 0.33$ and the existence of discontinuities in the slope of $\langle\Delta\rangle$ at the compositions $x = 0.33$ and 0.50. In the composition range $x > 0.50$, the $\langle\Delta\rangle$ values exhibit only a minor additional decrease in magnitude. In this region the rather large scatter in the values is primarily due to the fact that the doublet is only partially resolved. The composition $x = 0.17$ was studied with both a thick and a very thin absorber, with the hope that the thin absorber might lead to a reduction in the somewhat broadened linewidth. However, no significant improvement was found.

The Covalent Bond Length of Amorphous Te

The most Te rich amorphous composition investigated corresponded to $x = 0.075$. Smooth extrapolations of the $\langle\Delta\rangle$ and $\langle\delta\rangle$ values to $x = 0$ give

$$\Delta = 9.60 \pm 0.05 \text{ mm/sec}$$

$$\delta = +0.41 \pm 0.06 \text{ mm/sec}$$

These parameters are ascribed to amorphous Te. One notes that they are substantially different from those known [14] for crystalline Te:

$$\Delta = 7.74 \pm 0.04 \text{ mm/sec}$$

$$\delta = +0.50 \pm 0.03 \text{ mm/sec}$$

Since it is generally assumed that amorphous Te is predominantely two fold coordinated [15] as it is in the trigonal phase, it is necessary to explain this large difference in the extrapolated QS before attempting to advance a model describing the behavior of Ge_xTe_{1-x} as a function of x. We suggest that this phenomenon is related to the shorter covalent bond distance in amorphous Te.

It has been observed [16] that a linear relationship exists between the observed QS for Te($\underline{X}$) and R^{-3}, where R represents the length of the Te-X covalent bond and X represents orthorhombic S, trigonal Se, or trigonal Te. Figure 9 is reproduced from Ref. 16 and the straight line passing through the data represents the best fit to the QS as a function of R^{-3}.

In order to put amorphous points on Fig. 9, we need both Mössbauer QS measurements and direct measurements of the covalent bond distance in amorphous chalcogenides. Although no such Mössbauer measurements are yet available, indirect comparisons can be made with the amorphous compositions $Ge_{0.11}Te_{0.89}$ studied [17] by the X-ray radial distribution (RDF) technique and $Ge_{0.17}Te_{0.83}$ studied [7] with both X-ray and neutron RDF. Since these compositions are Te rich, and since the first-neighbor X-ray RDF peak is known [17] to be rather insensitive to composition, these comparisons should allow a good first approximation to amorphous Te. Accordingly, we have added to Fig. 9 our QS values for these compositions, with the first-neighbor RDF peak used to represent the covalent bond length. Two points are shown for the composition $Ge_{0.17}Te_{0.83}$ since the neutron and X-ray RDF give [7] peaks at 2.65 and 2.75 Å, respectively. This apparent discrepancy occurs because of the relative insensitivity to Ge in the X-ray technique leading to a heavier weighting of Te-Te

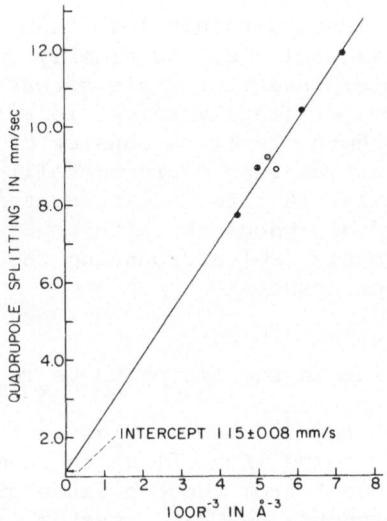

Fig. 9. Quadrupole splitting vs. R^{-3} where R represents the covalent bond length Te-X. Solid circles are for Te dissolved in crystalline hosts (Ref. 16). The amorphous points are for $Ge_{0.17}Te_{0.83}$ and $Ge_{0.11}Te_{0.89}$ and use for R the effective first neighbor distance from neutron RDF (circled cross) and X-ray RDF (circled dots). The errors in the QS measurements are less than the size of the circles.

bonds relative to the shorter Ge-Te bonds which are unresolved in both RDF spectra. The result appropriate to our observed Mössbauer QS values would be expected to be inbetween the two RDF values. In any case, we expect that the uncertainty in placing amorphous points on Fig. 9 will diminish when X-ray, neutron, and electron RDF data become available for amorphous Ge_xTe_{1-x} as a function of x.

We consider the data and the argument in Fig. 9 sufficiently convincing [18] to propose that our Mössbauer measurements indicate an effective covalent bond length of 2.62 Å for pure amorphous Te. However, it is not clear to us how this bond length, extrapolated from stabilized Ge_xTe_{1-x} films, may relate to the observed properties of the "amorphous" Te films prepared in the laboratory. For instance, a recent note [19] on the electron RDF of Te evaporated onto a substrate at liquid helium temperature shows a first near-neighbor position of 2.79 Å and suggests a two-fold coordinated chain length of about seven atoms.

The 2.79 Å bond distance, although less than the 2.8335 Å
characteristic of trigonal Te, is clearly inconsistent
with the current interpretation. This situation brings up
the crucial question, originally raised in Ref. 15, of
whether these "amorphous" Te films consist of randomly
separated chains, form as fine microcrystalline trigonal
crystals, or even exist in intermediate states having both
microcrystalline and amorphous character. Considering the
experimental uncertainty [20] surrounding the situation,
we leave the question open.

Structural Ordering in the Amorphous Ge_xTe_{1-x} System

We propose that the discontinuities in the slope of
the $<\Delta>$ versus x curve (Fig. 8) at the compositions
x = 0.33 and 0.50 result from the occurance of chemical
ordering in the amorphous Ge_xTe_{1-x} system. At each of
these compositions Te finds itself in an almost well defined local environment. Furthermore, we will show that
the QS at the composition x = 0.33 becomes more or less
definite on annealing the virgin amorphous films.

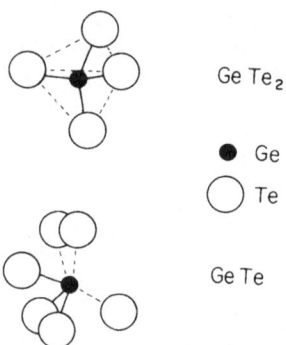

Ge Te$_2$

● Ge

○ Te

Ge Te

Fig. 10. Tetrahedral structure proposed in Ref. 21 for
amorphous GeTe$_2$ (above) and the black phosphorus structure
proposed in Ref. 22 for amorphous GeTe (below).

The smooth change in $\langle\Delta\rangle$ in the two composition ranges 0 to 0.33 and 0.33 to 0.50 originates, in our view, from a redistribution in the population of the inequivalent Te sites. It is also clear that one must invoke the presence of at least two inequivalent Te sites. In view of the simplicity of a structural model based on two inequivalent Te sites, it seems worthwhile to investigate its consequences.

At the composition $x = 0.33$ it is believed that GeTe$_2$ [21] exists in a structure similar to SiO$_2$. In this structure, the valence of each atom is locally satisfied in that each Ge is four-fold coordinated to Te, while each Te is two-fold coordinated to Ge. Figure 10 illustrates the GeTe$_2$ tetrahedron. At $x = 0.5$, it is believed that amorphous GeTe [22] exists in the black phosphorus structure. In this structure Te is coordinated to three first Ge neighbors and three second Ge neighbors (Fig. 10).

Two-Site Model of Amorphous Ge$_x$Te$_{1-x}$ $(0 < x < 0.5)$

In this model one assumes that the spectrum can be explained by two inequivalent Te sites. Since the linewidth of the components is comparable to the splitting between them, the two pairs of quadrupole doublets are not well resolved. The significant parameter is the average QS,

$$\langle\Delta\rangle = \sum_i N_i \Delta_i / \sum_i N_i \quad , \qquad (1)$$

where N_i and Δ_i designate the population and QS of the ith inequivalent site and any change in the isomer shift or in f factor between the two sites is assumed small.

If we assume the amorphous structure at $x = 0$ consists of predominantly two-fold coordinated Te, we can describe this structure as Te$_n$ chains where n is unknown. Similarly, the structure at $x = 0.33$ is GeTe$_2$ (perhaps the tetrahedra shown in Fig. 10, but this identification is in no way required). Then, in our two-site model, we invoke structural separation of the system into molecular GeTe$_2$ based and Te$_n$ based units. We leave open the question of the size of these units, but presumably they fall short of dimensions usually associated with a macroscopic phase separation. Thus for the composition range $0 < x < 0.33$

$$Ge_xTe_{1-x} = xGeTe_2 + \frac{1-3x}{n} Te_n \qquad (2)$$

Since each molecular unit of $GeTe_2$ has two equivalent Te sites, whereas Te_n has n equivalent Te sites, one may write

$$N_{i=1} = 2x$$

$$N_{i=2} = 1-3x$$

and this leads to

$$\langle\Delta\rangle = \frac{2x}{1-x} \Delta_{GeTe_2} + \frac{1-3x}{1-x} \Delta_{Te_n} \qquad (3)$$

where Δ_{GeTe_2} and Δ_{Te_n} designate the QS's at $x = 0.33$ and 0. Fit A in Fig. 11 is a plot of Eq. (3) keeping $\Delta_{GeTe_2} = 7.8$ mm/sec and $\Delta_{Te_n} = 9.6$ mm/sec. In fit B Δ_{GeTe_2} is changed to 8.04 mm/sec for the annealed films. These fits are rather encouraging.

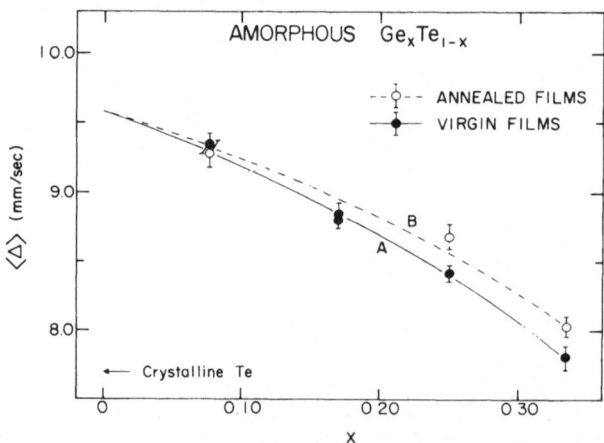

Fig. 11. Systematics of the mean quadrupole splitting $\langle\Delta\rangle$ for virgin and annealed films of Ge_xTe_{1-x} as a function of x for $0 < x < 0.33$. Curves A and B represent the two-site model constrained to agree with the data at $x = 0.075$ and $x = 0.33$.

In the composition range $0.33 < x < 0.50$, we again assume structural separation of the system into GeTe and GeTe$_2$ units, with all Te sites at $x = 0.33$ in a GeTe$_2$ surrounding and at $x = 0.5$ in a GeTe surrounding. Following the same development used above, we obtain:

$$<\Delta> = \frac{(3x-1)}{1-x} \Delta_{GeTe} + \frac{2(1-2x)}{1-x} \Delta_{GeTe_2} \tag{4}$$

This equation is plotted in Fig. 12. Curves A and B correspond to virgin and annealed films, respectively, with the values of Δ_{GeTe} and Δ_{GeTe_2} determined from measurements on virgin and annealed films of these compositions. The curves A and B do not describe the data well and curves C and D have been added to guide the eye. The nature of the discrepancy suggests that the failure of the two-site model in this region may be related to differences in the stabilities of the structures GeTe and GeTe$_2$ so that it is not permissible to treat them on an equal footing.

Effects of Annealing Amorphous Samples

Transport and optical properties of amorphous Ge$_x$Te$_{1-x}$ films are known to exhibit large annealing effects. For example, Rockstad and deNeufville [21] have observed an almost 15% increase in the conductivity activation energy of virgin GeTe$_2$ films on annealing. This and other changes [23] observed in the bulk properties of these films have been attributed to thermally induced structural changes. The Mössbauer effect, on the other hand, provides a simple and direct technique for observing structural changes on a microscopic scale.

For the purpose of annealing and crystallizing the amorphous films, a furnace was assembled which provided for supporting the films between two flat surfaces during heating. The virgin films were sandwiched between two Al blocks, one of which was heated by a 40 watt heating tape. This assembly about 1 in. in diameter and 2 in. long was supported in an evacuated bell jar. The temperature of the Al blocks was monitored with an Fe-constantan thermocouple and it could be regulated to $\pm$ 3°C by means of a variac.

Virgin GeTe$_2$ films were annealed at 175°C for one hour in vacuum. Mössbauer spectra of virgin and annealed films

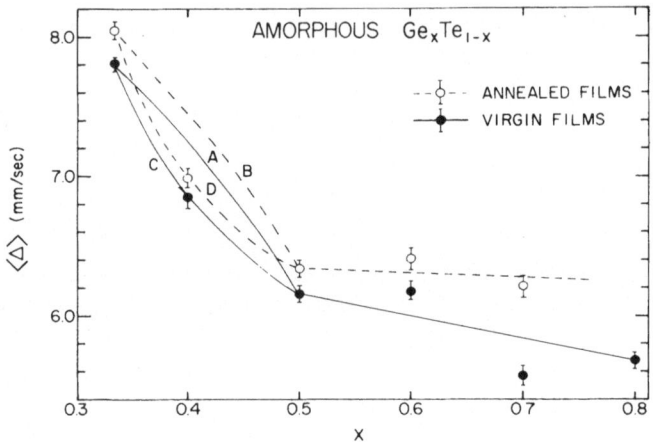

Fig. 12. Systematics of the mean quadrupole splitting
for virgin and annealed films of Ge_xTe_{1-x} as a function of x
for $0.33 < x < 0.50$. Curves A and B are the two-site model
constrained to agree with the data at $x = 0.33$ and $x = 0.50$.
Curves C and D are drawn to guide the eye.

were accumulated simultaneously in the experimental arrangement of Fig. 2. These measurements, summarized in Fig. 13,
were performed for several film thicknesses on both virgin
and annealed amorphous films. In both cases the linewidths
extrapolated to zero absorber thickness are larger than the
minimum observable linewidth $2\Gamma_n$. This suggests the presence
of a distribution of QS's in both of these non-crystalline
hosts. However, annealed $GeTe_2$ films exhibit an extrapolated linewidth which is approximately 0.4 mm/sec narrower
than that observed for virgin $GeTe_2$ films. Thus, the
process of annealing leads to a better defined QS. We also
note that the magnitude of the QS increases by 0.2 mm/sec on
annealing. The slope of the straight line obtained in a
plot of Γ_{exp} vs. film thickness is a measure of the absorber
f factor. The measurements in Fig. 13 indicate that the
process of annealing leads to a two-fold increase in the f
factor.

These large changes in the Mössbauer parameters on
annealing $GeTe_2$ reflect a predominantly nearest neighbor
effect. We suggest that the process of annealing leads to
a greater degree of structural order in the local $GeTe_2$
structure.

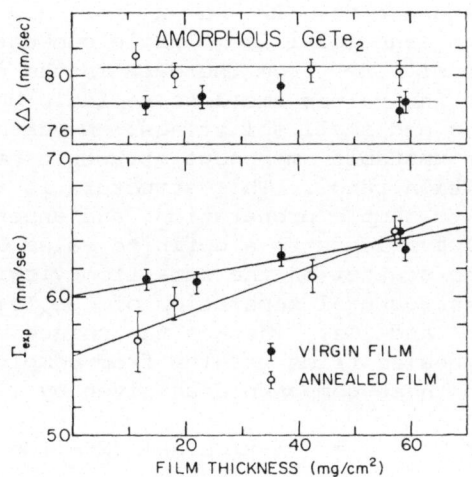

Fig. 13. Systematics of the mean quadrupole splitting $<\Delta>$ and the experimental linewidth Γ_{exp} as a function of film thickness for virgin and annealed films of amorphous GeTe$_2$. Note the two-fold increase in the f value of amorphous GeTe$_2$ on annealing. (The f value is proportional to the slope of the Γ_{exp} versus thickness curve.)

The excess linewidth of about 0.4 mm/sec above $2\Gamma_n$ observed for the quadrupole components of annealed GeTe$_2$ films raises some interesting questions. The possibility exists that the heat treatment used did not complete the process of structural ordering. An alternative explanation would attribute the broadening to the influence of more distant neighbors.

Measurements on annealed Ge$_x$Te$_{1-x}$ films were also performed at compositions other than GeTe$_2$. The $<\Delta>$'s obtained from these measurements are shown in Figs. 11 and 12 along with similar data on virgin films. Over a broad range of compositions annealed films exhibit a somewhat larger $<\Delta>$.

We note the absence of annealing effects in the most Te rich alloy studied (x = 0.075). This feature is not peculiar to Mössbauer measurements alone [21] and further suggests the presence of a low coordination Te site.

The independence of $<\Delta>$ on x for x > 0.5 shows
up better in the annealed films. While one can certainly
extract a value of $<\Delta>$ from the data in the range
0.5 < x < 0.7, the values are less reliable because of the
broad linewidths and small splittings. Nevertheless, we
believe than an unstable amorphous structure tends to occur
in this composition range. This structure is apparently
very sensitive to sample preparation, and annealing the
virgin films always leads to a definite value of $<\Delta>$. Thus,
we feel that the scatter in the data from virgin films reflects partial structural separation of Ge_xTe_{1-x} into
amorphous GeTe and Ge, whereas the unique value of $<\Delta>$
found in the annealed films results from complete structural
separation into these components as given by

$$Ge_xTe_{1-x} = (1-x)GeTe + (2x-1)Ge$$

Such a model would explain the independence of $<\Delta>$ and
the isomer shift $<\delta>$ on x.

Amorphous to Crystalline Phase Transformation

Amorphous Ge_xTe_{1-x} films are characterized by a well
defined glass transition temperature T_g, above which they
crystallize and exhibit long range order. This transition
is exothermic [24] and one can generally establish it by
measuring specific heats or by differential thermal analysis
[23]. The variation of T_g as a function of composition
for Ge_xTe_{1-x} has been studied by deNeufville [24] and
these systematics exhibit an anomaly at x = 0.33.

Spectra of crystalline Ge_xTe_{1-x} alloys for some compositions are reproduced in Figs. 5, 6 and 7. The crystalline films were obtained by heating the amorphous films to
a temperature 15°C above T_g in vacuum. At the composition
x = 0.075, the spectrum is characterized by a QS which
agrees well with that in Te metal. As the Ge content
increases, the spectral intensity at zero velocity builds
up until one reaches x = 0.5, where the spectrum acquires
a simple Lorentzian lineshape. The fits shown for spectra
at the compositions x = 0.075, x = 0.17, and 0.33 comprise
these lines: a quadrupole doublet having a QS and isomer
shift equal to that in Te metal and a single line at zero
velocity. The interpretation of these spectra thus places
Te in two inequivalent sites, one as in Te metal and the

other as in GeTe. Thus, the asymmetry of the spectrum at
x = 0.17 results from a positive isomer shift of the Te
metal doublet relative to the GeTe singlet. To summarize,
for x < 0.5 the amorphous to crystalline phase transition
in the Ge_xTe_{1-x} system leads to phase separation into
crystalline GeTe and Te metal:

$$a-Ge_xTe_{1-x} \rightarrow (1-2x)c-Te + x\ c-GeTe$$

where a and c stand for amorphous and crystalline.

Amorphous Te

The most Te rich amorphous Ge_xTe_{1-x} alloy investigated corresponded to x = 0.075. There is a practical
difficulty [15,19,20] in retaining the amorphous phase for
compositions x < 0.075 at room temperature, although it
may be possible to avoid crystallization with special
techniques. Mössbauer parameters for the extremum composition x = 0 are of particular interest as these correspond
to the amorphous phase of Te. One infers from the present
data that in amorphous Te, there exists a substantially
larger (24%) quadrupole splitting and a smaller isomer shift
than for the crystalline form. As discussed previously,
the data indicate that in the amorphous structure Te is
located in chemical surroundings that are quite different
from those of crystalline Te. Similar results are obtained
from NMR measurements [15] on rapidly solidified Te and
more recently from electron rdf results on evaporated Te
[19]. These authors have described amorphous Te to consist
of broken helical chains which average about six to seven
atoms long. We have shown that this picture in itself is
incomplete, and that a necessary ingredient of the structure
is that the bond length in the amorphous chains is shorter
than in crystalline Te.

Lucovsky and White [25] have recently advanced some
considerations based on chemical bonding and bond lengths
in amorphous semiconductors. They have stressed that the
chemical binding of second neighbors becomes progressively
stronger in going from crystalline S to Se to Te. In
trigonal Te, second neighbors are located in adjacent
helical chains. In amorphous Te the correlation between
chains is presumably destroyed. We feel this must affect
the geometry of the chains.

DISCUSSION AND SUMMARY

The structural picture [5] of amorphous Ge_xTe_{1-x} alloys that has emerged over the years from X-ray RDF, NMR and transport measurements, briefly is the following. Addition of Ge to amorphous Te leads to crosslinking of the Te chains and this process goes on until the composition $x = 0.33$ is reached where only Ge-Te bonds exist in the $GeTe_2$ structure. As this crosslinking process continues three types of Te environments emerge: One which has two Te near neighbors as in amorphous Te, a second which has one Te and one Ge near neighbor, and a third which has two Ge near neighbors as in $GeTe_2$. Formally, in our description of the average quadrupole splitting $<\Delta>$ as a function of composition in the range $0 < x < 0.33$, we have considered only two inequivalent sites. The simplicity of the model and the resulting fit to the data in this range might seem to justify our approach. However, more complicated approaches should also lead to reasonable agreement with the data for $0 < x < 0.33$.

In the composition range $0.33 < x < 0.5$, the inadequacy of the two-site model indicates that the GeTe and $GeTe_2$ structural units probably cannot be treated on an equal footing.

The isomer shifts of amorphous and crystalline GeTe were found to be nearly the same within the limits of experimental error. Indeed, recent XPS measurements of Shevchik et al. [8] show that the valence band structures of amorphous and crystalline GeTe are almost identical. We feel that these observations can be generally reconciled with the 3-fold coordination of GeTe in the black phosphorus structure, a model recently advanced by Bienenstock [22] to reinterpret the X-ray RDF measurements. A particularly attractive feature of this model is that Te is bonded to three first and three second neighbor Ge atoms in a coordination that is quite distorted from cubic symmetry. The process of crystallization of amorphous GeTe can be visualized as one that leads to a rearrangement of the first and second neighbors to form an almost octahedral coordination around Te. The preservation of coordination number in the two Te phases may be responsible for the unchanged isomer shift and valence band structure while the symmetry rearrangement leads to dramatic changes in the electric field gradient as observed.

Finally, the sizable linebroadening of the quadrupole components (Tables II, III) in amorphous GeTe persisted in annealed samples as well. Virgin GeTe films were annealed at 150°C for a 24-hour period. For this particular composition we are inclined to believe that the heat treatment used did not saturate the process of structural ordering. From a thermodynamic viewpoint, it would appear impossible to completely anneal samples at this composition since the annealing temperature T_g is estimated [18] to be about 50°C higher than T_x. We therefore associate the broad linewidths to a distribution of quadrupole splittings resulting from a certain degree of structural disorder.

ACKNOWLEDGEMENTS

We wish to thank Ron Nowicki for help in preparing the samples and Ned Dixon, Yusuf Mahmud, Art Wagner and Jim Oberschmidt for assistance in the experimental and computational work. In the earlier phase of these experiments, performed at the University of Cincinnati, the interest of Professor K. L. Chopra in making available some samples of GeTe is gratefully acknowledged. We are especially grateful to Professor Arthur Bienenstock of Stanford University for valuable discussions on the Ge_xTe_{1-x} system and to C. W. Seidel of New England Nuclear Corporation for assistance in some aspects of the source preparation. One of us (P.B.) would like to thank the nuclear group at Stanford University for the warm hospitality extended him during his stay.

REFERENCES

1. P. Boolchand, Nucl. Instr. and Meth. 114: 159 (1974).
2. J. Oberschmidt and P. Boolchand, Phys. Rev., in press.
3. C. Hohenemser and R. Rosner, Nucl. Phys. A109: 364 (1968).
4. E. Gerdau, W. Räth and H. Winkler, Z. Physik 257: 29 (1972).
5. N. F. Mott and E. A. Davis, Electronic Processes in Non-Crystalline Materials (Clarendon Press, Oxford, 1971).
6. F. Betts, A. Bienenstock, and S. R. Ovshinsky, J. Non-Cryst. Solids 4: 554 (1970), and references therein.
7. F. Betts, A. Bienenstock, D. T. Keating, and J. P. deNeufville, J. Non-Cryst. Solids 7: 417 (1972).

8. N. J. Shevchik, J. Tejeda, D. W. Langer, and M. Cardona, Phys. Rev. Letters $\underline{30}$: 659 (1973).
9. R. J. Blattner, L. K. Walford, and T. O. Baldwin, J. Appl. Phys. $\underline{43}$: 935 (1972).
10. A. Taylor and B. J. Kagle, Crystallographic Data on Metal and Alloy Structures (Dover, 1963).
11. C. N. King, W. A. Phillips, and J. P. deNeufville, Phys. Rev. Letters $\underline{32}$: 538 (1972).
12. D. Raj and S. P. Puri, Phys. Letters 29A: 510 (1969).
13. A. Wagner, MS Thesis, University of Cincinnati, 1974 (unpublished).
14. P. Boolchand, S. Jha, and B. L. Robinson, Phys. Rev. $\underline{B2}$: 3463 (1970).
15. A. Koma, O. Mizuno, and S. Tanaka, Phys. Stat. Solidi $\underline{B46}$: 225 (1971).
16. P. Boolchand, T. Henneberger, and J. Oberschmidt, Phys. Rev. Letters $\underline{30}$: 1292 (1973).
17. A. Bienenstock, F. Betts, and S. R. Ovshinsky, J. Non-Cryst. Solids $\underline{2}$: 347 (1970).
18. A. F. Wells, Structural Inorganic Chemistry (Clarendon Press, Oxford, 1962).
19. T. Ichikawa, J. Phys. Soc. Japan $\underline{33}$: 1729 (1972).
20. K. Bahadur and K. L. Chaudhary, Appl. Phys. Letters $\underline{15}$: 277 (1969).
21. H. K. Rockstad and J. P. de Neufville, 11th International Conference on Physics of Semiconductors (Polish Scientific Publishers, Warsaw, 1972) p. 68.
22. A. Bienenstock, J. Non-Cryst. Solids $\underline{11}$: 447 (1973).
23. R. K. Quinn and R. T. Johnson, Jr., J. Non-Cryst. Solids $\underline{12}$: 213 (1973).
24. J. P. deNeufville, J. Non-Cryst. Solids 8-10: 85 (1972).
25. G. Lucovsky and R. M. White, Phys. Rev. $\underline{B8}$: 660 (1973).

STUDY OF PHOTOCHROMIC STRONTIUM TITANATE BY MOSSBAUER EFFECT

P.J. Ouseph and C.T. Luiskutty

Physics Department, University of Louisville

Louisville, Kentucky 40208

Strontium titanate crystals, lightly doped with transition metal ions, show very interesting optical properties. Kiss and Faughnan discovered the photochromic properties of these crystals [1,2]. Blanc and Staebler [3] observed the electrochromic properties, that is, the color changes under the influence of electric fields. Also color changes can be produced by heat treatment [4]. Mossbauer studies reported here were conducted to understand the relationship between color changes and valance state of added impurities (Fe in our studies) in these crystals.

Crystals doped with iron and molybdenum (doubly-doped) are yellow in color and crystals doped with iron (singly-doped) are amber colored. As shown in Fig. 1 the colors of these crystals can be changed by the following methods.
1) Light irradiation (3900 - 4300 Å) of yellow crystals, doubly-doped crystals or singly-doped crystals made yellow by reduction, change their color to amber. On termination of light irradiation, the color changes back to yellow in $\sim$ 1 sec. 2) Heating crystals in air at 900°C changes the color of yellow crystals to amber and heating amber-colored crystals in partial vacuum ($\sim$ 1 Torr) also at 900°C changes their color to yellow. 3) Applying an electric field to singly-doped yellow crystals at 300°C for 1/2 hr. results in the formation of clearly defined yellow and amber colored regions, the yellow region being near the cathode and the darker region near the anode. If Fe-Mo doped crystals are used, one gets three regions: a dark region near the anode,

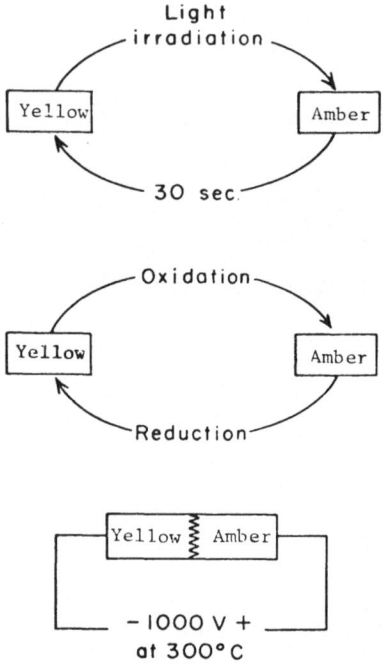

Fig. 1.   Processes that produce reversible color changes in SrTiO₃.

a yellow region in the center and a blue region near the cathode.

Electrical transfer processes involved in these color changes have been studied using optical [1,2], EPR [3] and other techniques [4].   The results of these studies are summarized here.   Iron enters SrTiO₃ lattice substitution-ally replacing Ti^{4+} (See Fig. 2), in high spin states.   If iron enters in an Fe^{4+} state, charge balance will be main-tained.   However, if it is in a lower ionic state in singly-doped crystals oxygen vacancy (V_O) and in doubly-doped crys-tals, the higher charge state of $Mo(Mo^{6+})$ is assumed to balance the charges.   Therefore, there will be five oxygen ions surrounding the iron in the 3+ state in singly-doped crystals, whereas there will be the regular number of six oxygen ions in the doubly-doped crystals.   The consequence

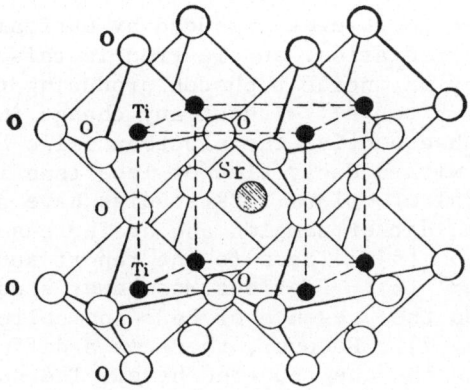

Fig. 2.   The $SrTiO_3$ structure, showing the octahedra of
oxygen atoms about the titanium atoms.   In Fe-substituted
crystal iron ions replaces the titanium ions.

of this difference on the IS is discussed later.   The optical absorption in the visible region is known to be very
small for Fe^{3+} [2] while the absorption is high for Fe^{4+}.
Hence it is assumed that the yellow samples contain Fe^{3+}
and the darker samples contain Fe^{4+}.   In the Fe and Mo crystals one expects Fe^{3+} and hence they are light yellow in
color while in Fe-doped crystals a larger concentration of
Fe^{4+} may account for the darker color.   On reducing the
sample, O^{2-} ions come out of the lattice, increasing the
oxygen vacancy and, consequently, increasing the Fe^{3+} content.   Crystals then become yellow in color.   On oxidation,
there is an increase in the Fe^{4+} content and also a color
change from yellow to amber.   When the electric field is
applied, O^{2-} ions move to the positive electrode side, increasing the Fe^{4+} concentration and, therefore, the optical
density.   Since O^{2-} is removed from the negative electrode,
the Fe^{3+} increases there making the region lighter.   When
oxygen is removed from the cathode side of Fe-Mo doped crystals, the charge state of Mo decreases from 6 to 5 and Mo^{5+}
is responsible for the blue color.   In the two methods discussed above, color changes are produced by increasing or
decreasing Fe^{4+} concentration as a result of changing the
oxygen ion concentration.   Therefore, these changes are more

stable compared to changes produced by optical irradiation.
During light irradiation, an electron is raised to the con-
duction band by an incident photon producing the valence
change from Fe^{3+} to Fe^{4+} without any change in the oxygen
vacancies. Subsequently, the electron falls back to the
iron ion site with a decay time of less than 1 sec producing the reversal of color. EPR studies have shown the presence of Fe^{3+} in darker samples and an increase in their number on reduction [5]. There is one report suggesting the
presence of Fe^{5+} [6]. Previous Mossbauer study by Bhide and
Basin indicated the presence of Fe^{3+} and colloidal iron in
these crystals [7]. However, there is a difference between
their study and the one reported here. The samples used in
this study are single crystals grown with impurities added
to the melt. Also it is well known that samples with higher
concentrations do not show any color switching properties.
Our studies, therefore, were made specifically to understand
the color switching properties of $SrTiO_3$ as a function of
the valence state of the iron ions.

Experimental

A Mossbauer spectrometer in constant acceleration mode
was employed. The samples to be studied were powdered and
used as absorbers in the experiments. The isomer shifts
were measured with respect to sodium nitroprusside standard.
Two crystals grown by the National Lead Company were used
for these studies. The two samples differed in their impurity concentrations: (1) yellow or light sample: 0.05 wt.%
Fe_2O_3, 0.2 wt.% MoO_3 plus 0.3 wt.% Al_2O_3; (2) amber colored
or dark samples: 0.1 wt.% Fe_2O_3. The iron oxide used in the
preparation of these crystals contained enriched iron, that
is 95% Fe^{57}.

For the low concentration Fe impurity in the amber-
colored crystal, we obtained the spectrum shown in Fig. 3(a).
This cannot be attributed to a single valence state of Fe
because the intensity of the two peaks are very different.
The peaks have isomer shifts 0.15 and 0.72 mm/sec. These
isomer shifts come in the range of those of Fe^{4+} and Fe^{3+}
respectively [8]. Thus at low concentrations of impurity,
a good fraction of Fe enters the strontium titanate lattice
as Fe^{4+}. A smaller amount of Fe ions are in the 3+ valence
state which has an associated oxygen vacancy to compensate
for the charge deficiency. The large halfwidth of Fe^{3+} peak

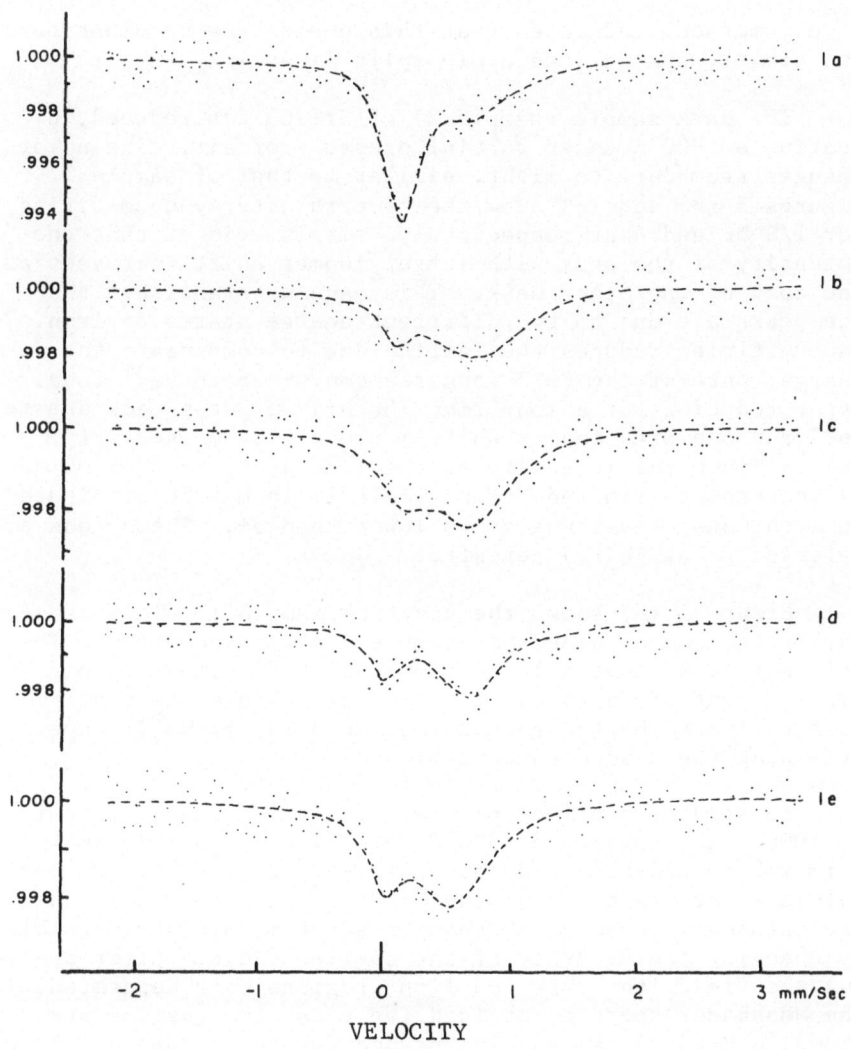

Fig. 3. Mössbauer spectra of (a) Fe-doped SrTiO₃, (b) Fe-doped SrTiO₃ after reducing for 1/2 hr, (c) Fe-doped SrTiO₃ after reducing for 4 hr, (d) Fe-doped SrTiO₃ after reducing for 36 hr, (e) Fe/Mo-doped SrTiO₃. The dashed lines are obtained by least square fit to Lorentzian using a computer. Isomer shifts are measured with respect to sodium nitroprusside standard. Total counts in each channel is ∿ 4 x 10⁶.

(~ 0.9 mm/sec) indicates that this could be a combination of two closely spaced quadrupole split lines.

The dark sample (sample 2) of $SrTiO_3$ was reduced, by heating at 900°C under partial pressure of air. The color changes from dark to light, similar to that of sample 1. Figures 3 (b) and (c) show the spectra after vacuum firing for 1/2 hr and 4 hr respectively. It is evident that the intensity of the peak with larger isomer shift increases at the cost of the other peak. This, again establishes that the peaks are due to two different charge states of iron. Vacuum firing reduces the lattice and to compensate the charge, part of the Fe^{4+} ions are converted to Fe^{3+} ions. After reducing for a long time (36 hr) we get a peak due to Fe^{3+} and one with isomer shift $\sim 0.0 \pm 0.03$ mm/sec (Fig. 3 (d)). Since the intensity of the line at 0.0 ± 0.03 mm/sec is increasing with reduction (Table 1) it may be attributed to iron ions of valence state lower than 3+. These ions are referred to as Fe(LV) hereafter.

Figure 3 (e) shows the spectrum sample 1. This is similar to the one obtained for sample 2 after reduction. In this sample we have a large amount of Fe^{3+} compared to a small amount of Fe(LV) state. The 3+ valence state of Fe is facilitated by the second ion, Mo being in Mo^{6+}, thus attaining the charge compensation.

To study the effect of electric field, a d.c. potential of 1000 V was applied at 300°C for 1/2 hr on a ~ 15 mm x 7 mm x 2 mm slightly reduced singly-doped crystal. We obtained a dark region near the anode and a light region near the cathode. These color changes are completely reversible by changing the polarity of the applied field. After applying the field, the dark and light regions were separated. The Mössbauer spectra obtained for these two regions are shown in Figs. 4 (a) and (b) respectively. Figure 4 (a) indicates a large amount of Fe^{4+} and a very small amount of Fe^{3+} in the sample. The ratio of areas of Mössbauer lines due to Fe^{3+} and Fe^{4+} is much smaller in this sample than in the original dark sample (Table 1). Figure 4 (b) shows that the region near the cathode contains mainly Fe^{3+} and Fe(LV). Partial reduction of the originally dark sample produces Fe^{3+} and Vo (oxygen vacancy) pairs. When electric field is applied O^{-2} accumulates near the anode and Fe^{3+} - Vo pairs near the cathode. In Fe/Mo doped crystals when oxygen is removed, the valence state of Mo decreased from 6 to 5.

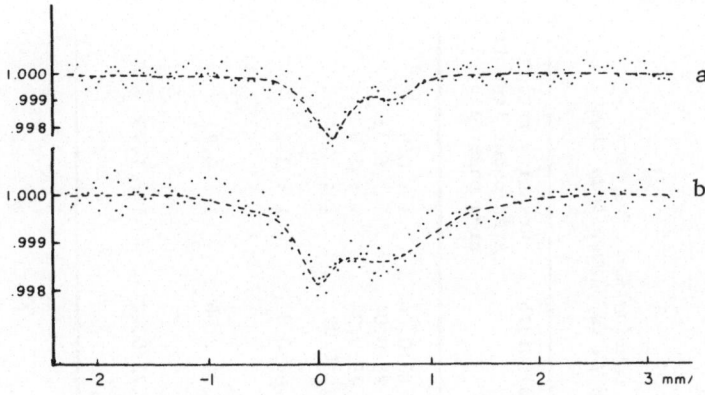

Fig. 4. Mössbauer spectra of the electrocolorised $SrTiO_3$ (Fe-doped). (a) dark region (near the anode). (b) clear region (near the cathode).

The Mo^{5+} is responsible for the blue color near the cathode.

Discussion

In the Mössbauer experiments, we found that, at low concentration of impurity iron, the valence state of the majority of Fe ions is 4+. On reduction of this sample, most of the Fe goes from Fe^{4+} to Fe^{3+} gradually and finally we get some Fe(LV) along with Fe^{3+}. The color changes from dark to light. When d.c. field is applied, Fe^{4+} ions gather near the anode and Fe^{3+} near the cathode producing dark and light colors near the anode and cathode respectively. In Fe/Mo doped $SrTiO_3$, 3+ is the predominant valence state of iron and the color is light (yellow). The ion Fe^{3+} does not absorb in the visible region and hence this accounts for the light color for the sample containing mainly Fe^{3+}. The ion Fe^{4+} has a large absorption cross section in the visible region;   this is the basis for the dark color of the Fe-doped $SrTiO_3$ and of the region near the anode when an electric field is applied. Our optical experiments with sample 1 and sample 2 also showed a larger absorption for the singly-doped sample (Fe) than for the multiply-doped sample (Fe-Mo) in the range of 4000 Å - 5000 Å.

Table 1. Mossbauer parameters of the different samples. Peak 1 of 1(c) is a combination of 2 peaks - one due to Fe^{4+} and the other a small one due to Fe(LV). Hence the halfwidth is larger than that of the peak 1 in other spectra.

	Absorption line 1 I.S. (mm/sec)	Halfwidth	Absorption line 2 I.S. (mm/sec)	Halfwidth	Intensity ratio Area under line 1: area under line 2
1(a) Fe-doped sample	$0.15\pm0.01(Fe^{4+})$	0.32 ± 0.04	$0.72\pm0.01(Fe^{3+})$	0.82 ± 0.04	1 : 0.9
1(b) Reduced 12 hr	$0.14\pm0.02(Fe^{4+})$	0.32 ± 0.04	$0.72\pm0.01(Fe^{3+})$	0.96 ± 0.04	1 : 6.23
1(c) Reduced 4 hr	$0.16\pm0.02(Fe^{4+})$	0.52 ± 0.06	$0.72\pm0.01(Fe^{3+})$	0.78 ± 0.02	1 : 2.6
1(d) Reduced 36 hr	$0.02\pm0.02(Fe_{LV})$	0.28 ± 0.02	$0.72\pm0.01(Fe^{3+})$	0.6 ± 0.02	1 : 3.42
1(e) Fe-Mo doped sample	$0.04\pm0.02(Fe_{LV})$	0.26 ± 0.02	$0.55\pm0.01(Fe^{3+})$	0.66 ± 0.02	1 : 3.7
2(a) Electrically colored sample - dark	$0.15\pm0.02(Fe^{4+})$	0.40 ± 04	$0.71\pm0.01(Fe^{3+})$	0.34 ± 0.06	1 : 0.27
2(b) Electrically colored sample - clear	$0.00\pm0.02(Fe_{LV})$	0.32 ± 0.02	$0.70\pm0.02(Fe^{3+})$	0.72 ± 0.12	1 : 1.75

It is appropriate at this point to discuss assignments of valence numbers to the iron ions in terms of their isomer shifts. Calculated Mössbauer parameters are given in Table 1. As mentioned earlier, the isomer shift of ions assigned to Fe^{3+} lies in the observed range of Fe^{3+}. For Fe^{4+}, the theoretically predicted value for the isomer shift is - 1.14 mm/sec with respect to sodium nitroprusside [9]. However, in the few compounds $(SrFeO_{2.3-5}$ and $La_{1-x}Sr_xFeO_3$ $(0 \leq x \leq 1))$ where Fe^{4+} has been observed [10-12] it is found that the isomer shift is in the range 0.2 to 0.35 mm/sec. Also recently during studies of photochromic TiO_2 where Fe replaces Ti, the presence of Fe^{4+} has been observed with the isomer shift ranging from -0.5 to +0.077 mm/sec [13]. Our results lie within the range of these results. Also the fact that the removal of oxygen decreases the amount of this species resulting in an increase in Fe^{3+} and the fact that the reverse effect occurs near the positive electrode where O^{2-} accumulates supports this assignment. All the light-colored samples Fe/Mo doped and reduced Fe-doped, show a peak, in addition to the broad Fe^{3+} peak, with an isomer shift close to -0.04 mm/sec. This inference is obvious from the results of the reduction data. On reducing the black sample for 1/2 hr, the ratio or Fe^{3+} to Fe^{4+} increases. Then this ratio decreases on further reduction indicating some ions going to an ionic state which has an isomer shift close to that of Fe^{4+}. Further, one can see the increases in the halfwidth of line 1 for the 4 hr reduced sample supporting the suggested explanation. After a long reduction (36 hr) the width of this line has gone down to a lower value and its position is clearly different from the original Fe^{4+} line. Bhide and Bhasin [7] have observed a line after reduction of their samples with low iron concentration   (0.3 at. wt.%) with an isomer shift of -0.3 mm/sec with reference to copper. This line they concluded arises from colloidal iron precipitate. They also have observed magnetic splitting of this line for higher concentrations. The line we are observing could be due to the same type of iron ions. However it is not possible to exclude the possibility that this line is due to low spin Fe^{2+}. Even though the Mössbauer spectra of the yellow samples look similar, the isomer shifts of Fe^{3+} of Fe/Mo doped samples is different from that of Fe-doped samples. This difference can be explained in terms of the changes in the overlap contribution to the electron density at the iron nucleus due to the changes in the number of oxygen ions in the two cases [14].

It was pointed out by Marshall [15] that the overlap of the wave functions representing the iron inner shells with ligand wave functions produces a significant change in the electron density at the nucleus. Overlap contributions to the electron density for Fe^{3+} have been calculated for Fe-0 as a function of the interatomic distance. The number of oxygen ions surrounding Fe^{3+} in strontium titanate could be 5 or 6 depending on the type of charge balance. The value of IS for Fe^{3+} in singly doped crystals is (+ 0.72 ± 0.02) mm/sec and for Fe^{3+} in doubly doped crystals is (+ 0.55 ± 0.01) mm/sec. As mentioned earlier these values are measured at room temperature with respect to sodium nitroprusside. The difference in the isomer shift is then (0.17 ± 0.03) mm/ sec.

The difference in the IS shifts may be written as

$$IS_{(Fe^{3+} - Mo^{6+})} - IS_{(Fe^{3+} - Vo)} = \alpha \Delta \rho (0).$$

where α is the calibration constant and $\Delta \rho (0)$ is the change in the electron density at the nucleus. The change in the electron density, $\Delta \rho (0)$, in this case is entirely due to the change in the overlap contribution due to the removal of one oxygen ion. The overlap contribution due to the oxygen ions (six ions in the octahedra) has been calculated as a function of the Fe-0 distance [16]. The Fe-0 distance in $SrTiO_3$ is 1.95 Å, and, therefore, from [16], one gets $1/6 \times 3.63 \ a_0^{-3}$ = $0.605 \ a_0^{-3}$ for $\Delta \rho (0)$. By using the known value of α, which is $-(0.316 \pm 0.011) a_0^3$ mm/sec [17], $IS_{Fe^{3+} - Mo^{6+}} - IS_{Fe^{3+} - V_0}$ = -0.191 ± 0.007 mm/sec. This value agrees with the experimental value within the limits of the error. In a previous study to determine the effect of overlap contribution, Šimánek and Šroubek [16] measured IS of divalent iron in CoO as a function of pressure. They were able to attribute the change in IS to changes in overlap contribution only after assuming a small value for $\alpha(-0.1147 \ a_0^3)$.

These experiments show that the changes in optical density are due to changes in the amount of Fe^{4+}, brought about by the removal of oxygen. Removal of oxygen by reduction and electric treatment produces permanent color changes. On the other hand, changes in ionic state from Fe^{3+} to Fe^{4+} resulting from light irradiation have a short decay time, typically 30 sec. Consequently, the color changes in the latte case reverse rapidly. Also these studies support the view that oxygen vacancy in singly-doped crystals and higher valence states of Mo provide the charge balance in lighter crystals containing iron ions of valence state 3 or lower.

Acknowledgements

This work is partly supported by grants from NSF (No. GH-35864) and from the Research Corporation. The authors would like to acknowledge many helpful discussions with Dr. I. Dezsi, Hungarian Academy of Sciences and Dr. B.W. Faughnan of RCA Laboratories, Princeton, New Jersey. We are grateful to Mr. H. Groskreutz for his help in preparing the computer program.

References

1. B.W. Faughnan and Z.J. Kiss, Phys. Rev. Letters $\underline{21}$, 1331 (1968).
2. B.W. Faughnan and Z.J. Kiss, IEEE, J.Q.E. $\underline{QE-5}$, 17(1969).
3. J. Blanc and D.L. Satebler, Phys. Rev. $\underline{B4}$, 3623 (1971).
4. R.L. Wild, E.M. Rocker and J.C. Smith, Phys. Rev. $\underline{B8}$, 3828 (1973).
5. B.W. Faughnan, Phys. Rev. $\underline{B4}$, 3623 (1971).
6. K.A. Müller, Th. Von Waldkirch, W. Berlinger and B.W. Faughnan, Solid State Communications $\underline{9}$, 1097 (1971).
7. V.G. Bhide and H.C. Bhasin, Phys. Rev. $\underline{172}$, 290 (1968).
8. C.T. Luiskutty and P.J. Ouseph, Solid State Communications $\underline{13}$, 405 (1973).
9. L.R. Walker, G.K. Wertheim, and V. Jaccarino, Phys. Rev. Letters, $\underline{6}$, 98 (1961).
10. P.K. Gallagher, J.B. Macchesney and D.N.E. Buchannan, J. Chem. Phys. $\underline{41}$, 2429 (1964).
11. G. Shirane, D.E. Cox and R.L. Ruby, Phys. Rev. $\underline{125}$, 1158 (1962).
12. U. Shimony and J.M. Knudsen, Phys. Rev. $\underline{144}$, 361 (1966).
13. A.J. Nozik, J. Phys. C. Solid State Phys. $\underline{5}$, 3147 (1972).
14. C.T. Luiskutty and P.J. Ouseph, Phys. Stat. Sol. $\underline{(b)58}$, K171 (1973).
15. W. Marshal, Proc. II Internat. Conf. on the Mössbauer Effect, Ed. A.H. Schoen and D.M.J. Compton, Wiley, New York 1962 (p. 263).
16. E. Šimánek and Z. Sroubek, Phys. Rev. $\underline{163}$, 275 (1967).
17. H. McKlitz and P.H. Barret, Phys. Rev. Letters $\underline{28}$, 1547 (1972).

LATTICE SOFTENING AND PHONON MOMENTS IN HIGH T_c SUPERCONDUCTORS[†]

C.W. Kimball, S.P. Taneja and L. Weber
Northern Illinois University
DeKalb, Illinois 60115
    and
F.Y. Fradin
Argonne National Laboratory
Argonne, Illinois 60439

ABSTRACT

Mössbauer experiments are described which were designed to contrast the lattice properties of high and low T_c superconductors having the β-W structure. The technique of Kitchens, Craig and Taylor was employed to obtain the ratios of phonon moments which are of importance in McMillans' analysis of 'strong-coupled' superconductors. The factor $1/(<\omega>_0/<\omega^{-1}>_0)$ corresponds to the phonon-only part of the electron-phonon coupling parameter λ and, hence, is related to T_c in the theory. Mössbauer results for the $V_3Ga(T_c \sim 15°K)$-$V_3Sn(T_c \sim 4°K)$ pseudobinary compounds indicate that the concentration dependence of the factor $1/(<\omega>_0/<\omega^{-1}>_0)$ is weaker than that expected from T_c. Lattice softening was observed in the Ga-rich compound, and in a quasiharmonic approximation the temperature dependence of the Debye temperature was obtained.

INTRODUCTION

Many high temperature superconductors are binary or ternary compounds with the β-W (A15) crystal structure.

[†]Based on work performed under the auspices of the National Science Foundation and the U.S. Atomic Energy Commission.

These compounds exhibit anomolous physical properties, e.g., in their low temperature specific heat, susceptibility and elastic stiffness.[1] The most salient features of the properties of these compounds for this discussion are a high density of states near the Fermi energy and the softening of the lattice at low temperature. In some β-W compounds with high transition temperature T_C a phase transition from a cubic to tetragonal unit cell occurs above T_C. Moreover, compounds with high T_C tend to exhibit large anharmonic-related behavior. The high T_C in the β-W compounds has been attributed by some investigators to a high peak in the density of states near the Fermi energy E_F (and a consequent low effective Fermi temperature) and by others to lattice properties, e.g., mode softening and large anharmonicity.[1]

The Debye-Waller factor obtained from Mössbauer results at ^{119}Sn sites in Nb_3Sn (T_C=18°K) indicates that Sn is behaving very anharmonically and that the Debye-temperature is very low[2]. But, x-ray measurement of the Debye-Waller factors for both constituents in Nb_3Sn indicates that the temperature dependence of the lattice behavior can be described by a Debye model (that is, the lattice is behaving harmonically)[3]. The Nb_3Sn compound can undergo mode softening with or without a phase transition.

A Mössbauer study has been undertaken to contrast the lattice properties of high and low T_C superconducting compounds with the β-W structure. Mössbauer results have been obtained for the V_3Ga(T_C∿15°K)-V_3Sn(T_C≈4°K) system. Specific heat and susceptibility measurements for the samples used in the Mössbauer investigation do not indicate a phase transition in this system.[4]

BACKGROUND

The area of Mössbauer absorption is related to the Debye-Waller factor in a harmonic solid, that is,

$$-\ln A \; \alpha \; -\ln f \; ,$$

where f is the Mössbauer fraction. Moreover,[4]

$$-\ln f = <x^2> \, K^2 \; ,$$

where K is 2π divided by the wave-length of the gamma ray.

In the harmonic approximation

$$\langle x_j^2 \rangle = \hbar \, \langle \omega^{-1} \rangle \, / M_j \; ,$$

where $\langle x_j^2 \rangle$ is the mean squared displacement of the jth atom in the unit cell, M_j is the mass of the Mössbauer atom, and $\langle \omega^{-1} \rangle$ is the (-1) moment of the phonon spectral density as defined below. For a Debye solid

$$\langle x_j^2 \rangle = \frac{3\hbar^2}{M_j k_B \theta_D} \left[\frac{1}{4} + \left(\frac{T}{\theta_D} \right)^2 \int \frac{x dx}{e^x - 1} \right]$$

and for $T > \theta_D$

$$\langle x_j^2 \rangle = \frac{3\hbar^2}{M_j k_B} \left(\frac{T}{\theta_D^2} \right) + \langle x_j^2 \rangle \, zp \; ,$$

where $\langle x_j^2 \rangle$ zp is the zero-point motion. The mean-squared displacement is linear in temperature at high temperature with a slope proportional to θ_D^{-2}. Thus, at high temperature the slope of $-\ln A$ vs T is very sensitive to the value of θ_D.

In the experiments described here no correction has been made for background radiation and, consequently, only a relative value of the Debye-Waller factor is obtained. However, the Debye temperature can be obtained from the temperature dependence of $-\ln A$.

The position of the centroid of the Mössbauer pattern $\delta(T)$ is determined by two factors:

$$\delta(T) = \delta_{IS}(T) + \delta_{th}(T) \; ,$$

where $\delta_{IS}(T)$, the isomer shift, is related to the electrostatic energy of the interaction between the electronic and nuclear charge and $\delta_{th}(T)$, the thermal shift, is related to the lattice energy. The thermal shift is proportional to the mean-squared velocity of the Mössbauer atom

$$\delta_{th}(T) = \langle v^2 \rangle / 2c \qquad .$$

In the harmonic approximation,

$$\langle v^2 \rangle = \frac{3\hbar}{M_j} \langle \omega \rangle \; ,$$

where $<\omega>$ is the first moment of the phonon spectral density. For a Debye solid

$$<v_j^2> \ = \ <v_j^2> \ zp \ + \ \frac{9k_BT^4}{\theta_D^3 M_j} \int^{\theta_D/T} \frac{x^3 dx}{e^x-1} \ \ ,$$

where

$$<v_j^2> \ zp \ = \ 9k_B\theta_D/8M_j \ \ .$$

The thermal shift $\delta_{th}(T)$ is given here to within a constant because only the difference between source and absorber energy is measured in the Mössbauer experiment. Note that $\delta_{th}(T) - \delta_{th}(0)$ is a measure of the change in thermal energy in the harmonic approximation and determines the Debye temperature and the change in thermal energy. At high temperature $(T > \theta_D)$,

$$<v_j^2> \ = \ 3 \ k_B \ T/M_j$$

and is independent not only of the Debye temperature but also of the thermal model of the solid.

The magnitude of the isomer shift is proportional to the product of a constant nuclear factor and the electron charge density at the nucleus. For ^{119}Sn the nuclear factor is positive. A change in isomer shift with temperature can result from a change in electron character due to the thermal smearing of the Fermi surface with concomitant differences in band occupation and also from volume changes due to thermal expansion.

Kitchens, Craig and Taylor[6] have reviewed their early work in which they show that a plot of $-\ln f(T)$ vs $\delta(T)$, yields a curve which defines a parameter $S(T)$ that is given in the harmonic approximation by

$$S(T) \ \simeq \ -\ln f(T)/\delta_{th}(T) \ \sim \ <\omega>/<\omega^{-1}> \ \ ,$$

where $<\omega^\ell> \ = \ \int F(\omega)(\bar{n}+\frac{1}{2}) \ \omega^\ell \ d\omega$ (normalized); $F(\omega)$ is the phonon spectral density and $(\bar{n}+\frac{1}{2})$ is the Bose kernel. In McMillan's application of the BCS theory to strong-coupled $(\lambda \approx 1)$ superconductors, the transition temperature is given by

$$T_c \ = \ \frac{\theta_D}{1.45} \ \exp\left[-\frac{1.04(1+\lambda)}{\lambda-\mu^*(1+0.6\lambda)}\right] \ \ ,$$

where μ^* is a Coulomb pseudo-potential and λ is the electron-phonon coupling parameter

$$\lambda = \frac{N(0)<I^2>}{M<\omega>_0/<\omega^{-1}>_0} \quad .$$

$N(0)$ is the bare density of electron states, $<I^2>$ is the electron-ion scattering matrix element averaged over the Fermi surface, M the atomic mass and $<\omega^\ell>_0$ is the ℓth moment of the phonon distribution $F(\omega)$ at $T = 0°K$. McMillan contended that, for $N(0)$ sufficiently high, $N(0)<I^2>$ is approximately a constant and that the phonon properties are the principal determinant of the magnitude of T_c. In principle, the phonon moments of importance in McMillan's treatment of strong-coupled superconductors can be obtained from Mössbauer measurements.

The validity of the assumptions in McMillan's treatment of strong-coupled superconductors rests on the relationship of $F(\omega)$ to the product $\alpha^2(\omega)F(\omega)$, where $\alpha^2(\omega)$ is the average electron-phonon interaction. McMillan assumes that $\alpha^2(\omega)$ is almost frequency independent and obtains the above expression for λ. There are plausible physical arguments for making this assumption for the compounds studied here; namely that neutron inelastic scattering and tunneling measurements on Pb, a three-dimensional 'soft' phonon system, indicate that the frequency dependence of $\alpha^2(\omega)F(\omega)$ and $F(\omega)$ are similar.[8] However, in low dimensional solids larger differences between $\alpha^2(\omega)F(\omega)$ and $F(\omega)$ are expected. Neutron scattering measurements of the phonon dispersion relations for the β-W compound Nb_3Sn, a high T_c superconductor, show phonon softening over a large range of q; this behavior is not typical of a Kohn anomaly in a low-dimensional solid[1].

EXPERIMENTAL

In addition to requiring care in the measurement of intensities, this study also required accurate determination of the shift at low temperature where the lattice contribution is in the T^3 region. Consequently, the absorber and source temperatures had to be fixed very precisely. For the experiments in the temperature range from 2 to 300°K the linear motor and sources were installed in a chamber consisting of two massive thermal shields; the

outermost shield was completely surrounded by an ice-water
mixture; the chamber was filled with dry nitrogen at a
pressure of $\sim$ 10 mm of Hg. Aluminized mylar windows are
used throughout. The temperature of the BaSnO$_3$ (^{119}Sn),
source, Fe (^{57}Co)-in-iron source and iron absorber (5-9's
annealed, natural iron foil, 0.5 mil thick) was monitored
by thermocouples. It was found that no further control was
necessary as the probe-source, and calibration-source
and absorber, remained at 273.56 ± 0.03°K. The zero of
velocity and the velocity calibration measured with the
^{57}Co were determined simultaneously with the measurement
of the Mössbauer parameters of the ^{119}Sn in the superconducting compounds. Over four and eight month time intervals repeated measurements of the temperature dependence
of the low temperature shift (T < 120°K) for V$_3$Sn$_{0.1}$Ga$_{0.9}$
yielded values which varied by ± 0.002 mm/sec at a given
temperature; Γ was 0.9 mm/sec and the total count was $\sim 10^6$ per
channel (folded).

A palladium foil of 2 mil thickness was used as a
critical absorber. Ruby[9] has studied the effect of various
thicknesses of Pd on the attenuation of the 23.88 keV γ-ray
and the adjacent x-rays, and also various geometries on the
efficiency of detection. Figure 1 shows the effect of a
2 mil Pd absorber in the geometry of our experiment. The

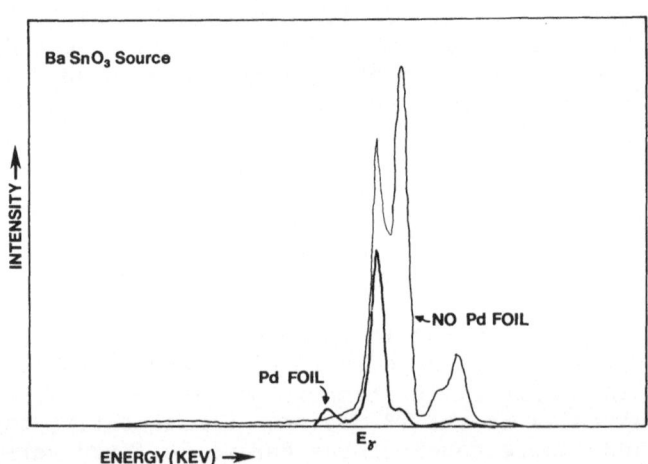

Fig. 1 Effect of 2 mil Pd critical absorber on ^{119}Sn x-rays
and γ-ray (23.88 keV), (Ruby[9]).

detector is a 300 mm^2 x 5 mm thick Si(Li) solid state detector with $\sim$ 400 eV resolution (FWHM) at 24 keV. The Pd
critical absorber is placed approximately midway between
the source and the detector. Placing the absorber close to
the source increases the production of Pd x-rays and placing
it close to the detector enhances the efficiency of detecting Pd x-rays. A proportional counter was used as the
detector in our experiments.

The line width (FWHM) of a thin natural iron absorber
placed in the sample holder in the Dewar, i.e., held in the
same manner as the $V_3Ga_xSn_{1-x}$ compounds are, was 0.201 mm/sec
($\pm$ 0.004). The line width (FWHM) of a thin $BaSnO_3$ absorber in the sample holder was 0.92 mm/sec for a 15 mC
$Ba^{119}SnO_3$ source. Figure 2 shows a typical absorption plot
at 4.4°K and 300°K for $V_3Sn_{.1}Ga_{.9}$.

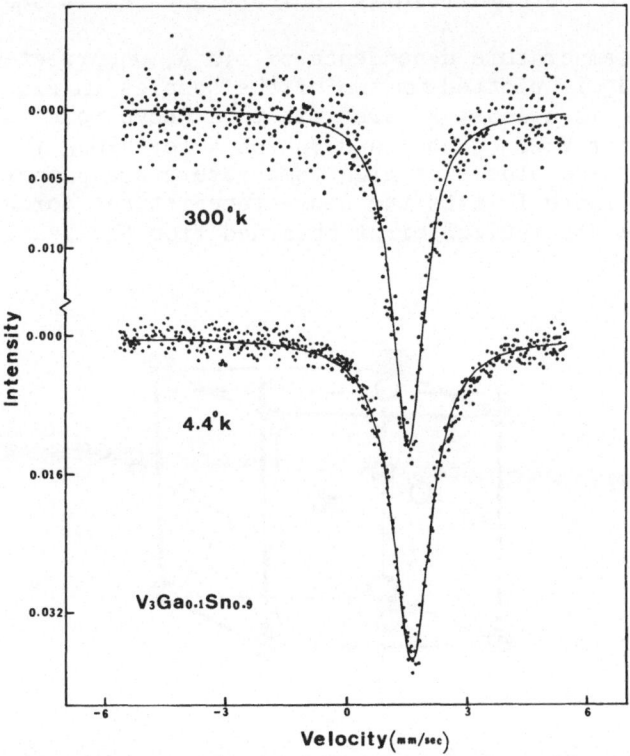

Fig. 2 Mössbauer absorption for $V_3Ga_{0.1}Sn_{0.9}$ absorber at
4.4°K and 300°K.

RESULTS AND DISCUSSION

Mössbauer measurements were made at ^{119}Sn nuclei in the β-W compounds $V_3Ga_{0.9}Sn_{0.1}$ ($T_c \sim 11°K$), $V_3Ga_{0.5}Sn_{0.5}$ ($T_c \sim 6°K$) and $V_3Ga_{0.1}Sn_{0.9}$ ($T_c \sim 4°K$) in the temperature range from 2.2°K to 300°K. Measurements were also made for the $V_3Ga_{0.9}Sn_{0.1}$ and $V_3Ga_{0.1}Sn_{0.9}$ compounds from 300°K to 900°K in a second spectrometer in which the source temperature was not clamped.

The Sn and Ga atoms are randomly distributed on a bcc lattice; the V atoms occur on the bcc faces and extend throughout the crystal in linear chains with the chains in adjacent cube faces perpendicular to one-another (see Fig. 3)[10]. The near-neighbors of the Sn atoms are V atoms (12) and the local symmetry is cubic. There can be a finite field gradient (EFG) arising from long range disorder on the bcc lattice (but nmr results indicate any EFG is small[13]).

The temperature dependence of -ln A, uncorrected for background, is plotted for two of the samples in Fig. 4. The solid line is a least-squares fit to a Debye model with $\theta_D = 269°K$ for $V_3Ga_{0.9}Sn_{0.1}$ and $\theta_D = 266°K$ for $V_3Ga_{0.1}Sn_{0.9}$ (see Table I). The slopes at high temperature are proportional to $1/\theta^2$. Table I lists the Debye temperatures for all samples and the statistical error obtained from the least-squares

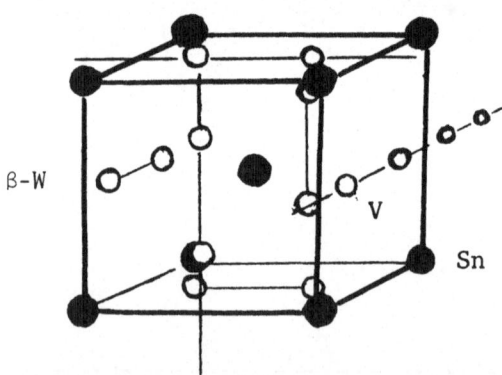

Fig. 3 Unit cell of β-W compound V_3Sn. The Sn atoms form a bcc lattice. The V atoms are in perpendicular linear chains with two V atoms in each face of the unit cell.

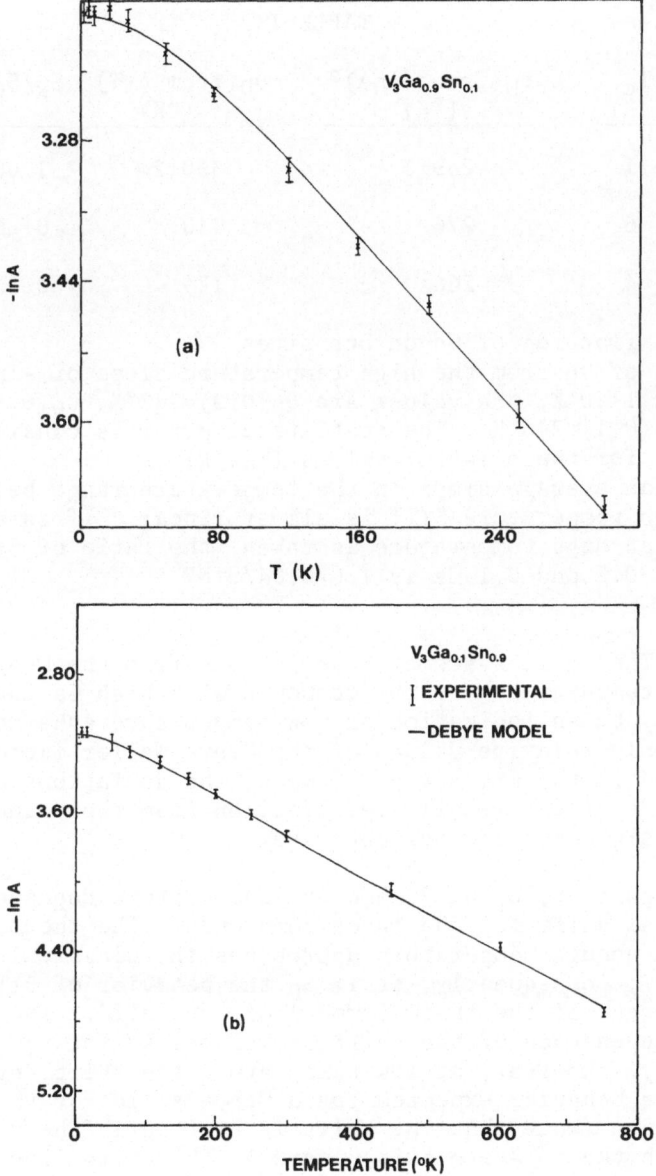

Fig. 4   (a) Temperature dependence of -ln A for $V_3Ga_{0.9}Sn_{0.1}$.
The solid line is a Debye curve for $<x^2>$ with $\theta_D=269°K$.
(b) Temperature dependence of -ln A for $V_3Ga_{0.1}Sn_{0.9}$.   The
solid line is a Debye curve for $<x^2>$ with $\theta_D=264°K$.

TABLE I

x^a	T_c (°K)	θ_D(from $-\ln A$)[b] (°K)	θ_D(from $\delta(T)$) (°K)	$S_x/S_{0.9}$[c]
0.9	11	269±3	350±24	1.00
0.5	6	276	340	0.86
0.1	4	266	174	0.23

[a]atomic fraction of Ga on bcc sites
[b]Values of θ_D from the high temperature slope of $-\ln A$ vs T; for T<50°K, the values are $\theta_D(0.9)$=242°K, $\theta_D(0.5)$=263°K and $\theta_D(0.1)$=246°K. The statistical error is considerably larger for these latter values (±20°K).
[c]S_x is an average slope in the temperature range below 50°K. At high temperature $S(T)$ is almost linear. If an average value at high temperature is taken, the ratio of $S(T)$ for x=0.9, 0.5 and 0.1 Ga is 1.0/0.94/0.87.

fits. The small systematic deviations from the Debye curve at low temperature for the compound with high Ga concentration may be an indication of low temperature anharmonicity. Since only relative values of the Debye-Waller factor are measured in the present experiment, the deviations of the magnitude of the recoil-free fraction from the values for a harmonic model are not obtained.

Figures 5, 6 and 7 show the temperature dependence of the total shift for the three compounds. The thermal shift at high enough temperature approaches the classical limit $(3/2)k_BT$; consequently, at $T>>\theta_D$ the behavior of $\theta(T)$ is independent of the thermal model of the solid. The temperature dependence of the shift is Debye-like for $V_3Ga_{0.1}Sn_{0.9}$ (Fig. 5). However, at low temperature the shift deviates from the behavior expected for a Debye solid for the 0.9 Ga compound. (Note that the shift is plotted in the usual manner with the negative axis upward.) The solid lines are the thermal energy as a function of T on the basis of Debye model fits to the 4.2°K point and the high temperature data. Note the large difference between the values of θ_D derived from the $-\ln A$ data and those derived from the thermal shift data (Table I). The different Debye temperatures result from the fact that the two Mössbauer

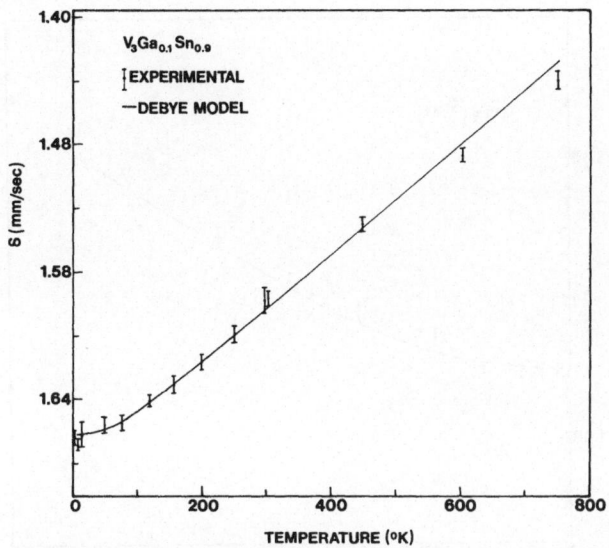

Fig. 5   Temperature dependence of $\delta(T)$ for $V_3Ga_{0.1}Sn_{0.9}$.
The solid line is a Debye curve for $<v^2>$ with $\theta_D = 174°K$.

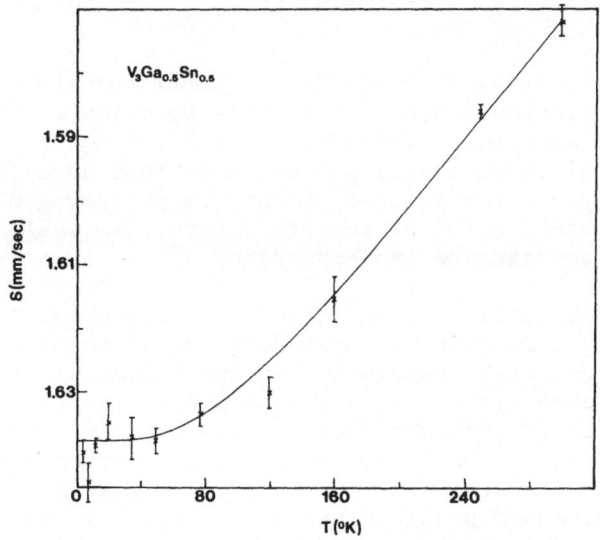

Fig. 6   Temperature of $\delta(T)$ for $V_3Ga_{0.5}Sn0.5$.   The solid
line is a Debye curve for $<v^2>$ with $\theta_D = 340°K$.

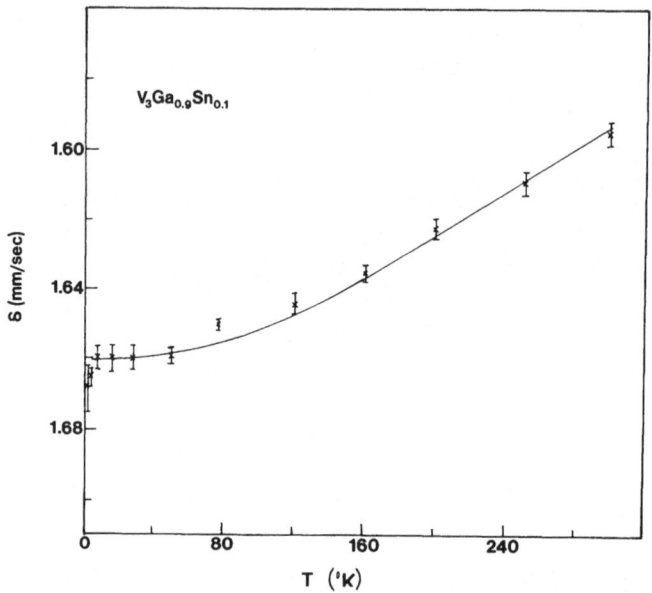

Fig. 7 Temperature dependence of $\delta(T)$ for $V_3Ga_{0.9}Sn_{0.1}$.
The solid line is a Debye curve for $\langle v^2 \rangle$ with $\theta_D = 350°K$.

parameters are related to different moments of the 'real'
phonon distribution which is not well-approximated by a
single Debye spectrum. Therefore, the Debye model which
describes a given Mössbauer parameter is that necessary
to properly yield the related phonon moment averaged over
the 'real' phonon spectral density. Later, we shall use
this result to describe the dependence of -ln A on $\delta(T)$.

The isomer shift can be temperature dependent through
a change in volume with temperature (thermal expansion) and
through changes with temperature in the character of the
occupied valence states. The pressure experiments on Sn
in various solids indicate that the isomer shift decreases
with decreasing volume; that is, the total charge density
at the nucleus decreases with decreasing volume.[11] The
effect of increased p-shielding apparently overcomes the
tendency towards an increase in s-like density at the nucleus. The thermal coefficient of expansion α for the β-W
compounds is quite small below 40°K; α is positive at

higher temperatures.[12] Therefore, as T increases from
temperatures near 40°K the solid expands and the charge
density at the nucleus should increase. An increase in
isomer shift should result. The expected change in isomer
shift with temperature is opposite to the total change in
shift observed for $V_3Ga_{0.9}Sn_{0.1}$.

A density-of-states function which accounts for the
temperature dependence of the susceptibility and spin lattice relaxation rate of ^{51}V in V_3Ga-V_3Sn yields a chemical
potential that increases with increasing temperature.[13]
An upward shift of μ with T leads to an increase in s-
occupation and a decrease (by the same number) of the d-
occupation. Both of these effects also lead to an increase
in isomer shift with increasing temperature. Hence, any
change in isomer shift with temperature will be opposite
in sign to that of the observed change in shift for
$V_3Ga_{0.9}Sn_{0.1}$. We conclude that the observed variation of
shift with temperature in the high T_C compound cannot be
due to an isomer shift. (We note that the behavior of the
isomer shift is more difficult to predict for compounds in
which a phase transition occurs, such as Nb_3Sn). Moreover,
the estimated change in s-occupation of $\sim 3(10^{-3})$ leads to
a change in shift which may be as large as ± 0.005 mm/sec.
We therefore may be underestimating the change in Debye
temperature (see Fig. 8 and Fig. 9).

There is ample experimental evidence that the high T_C
compounds with the β-W structure tend to undergo lattice
softening with decreasing temperature, and may undergo
phase transitions. A most striking example of lattice
softening is taken from the measurement of the elastic properties of V_3Si by Testardi and Bateman[14] shown in Fig. 8(a).
Testardi et al.[15] find a large temperature dependent softening in V_3Ga as well (see Fig. 8(b)). We attribute the non-
Debye like behavior of the shift in $V_3Ga_{0.9}Sn_{0.1}$ to lattice
softening, i.e., to a temperature dependence of the Debye
temperature, and treat the thermal shift in a quasiharmonic
approximation. A thermal shift for each temperature* below
160°K is combined with shift data for high temperature.
For T>200° $\delta(T)$ vs T is approaching classical behavior.
Using the Debye function for the thermal energy, a Debye

---

*To reduce statistical error, the low temperature data
points are combined in pairs.

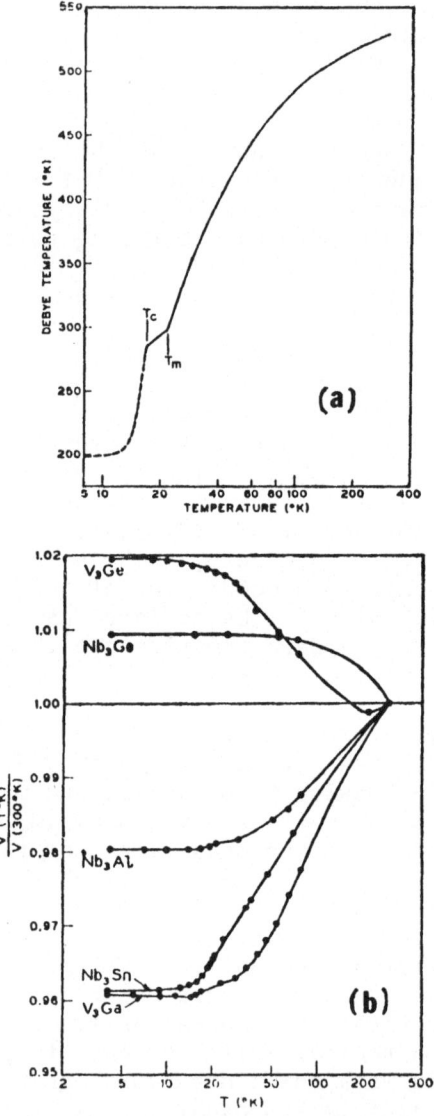

Fig. 8 (a) Temperature dependence of the Debye temperature calculated from elastic moduli for a single crystal of V_3Si (From Testardi and Bateman[14].) (b) Temperature dependence of the velocity of sound in β-W compounds (From Testardi et al.[15])

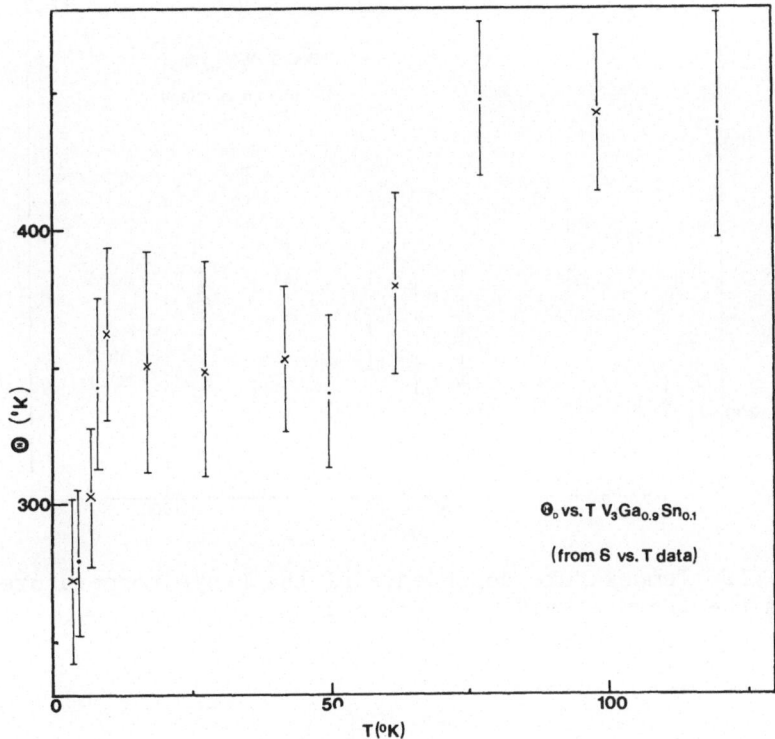

Fig. 9 Temperature dependence of the Debye temperature
from thermal shift for $V_3Ga_{0.9}Sn_{0.1}$.

temperature is obtained for each $\delta(T)$. This procedure is
equivalent to generating a family of Debye curves for different θ_D and treating the θ_D as a temperature dependent
quantity (quasiharmonic approximation). Each experimental
$\delta_{th}(T)$ for T<160 falls on a Debye curve with a different
characteristic temperature. Fig. 9 shows θ_D vs T for
$V_3Ga_{0.9}Sn_{0.1}$ and Fig. 10 for $V_3Ga_{0.5}Sn_{0.5}$. A softening of
the lattice is clearly observed between 80°K and 50°K for
the high T_c, 0.9 Ga compound.

Figure 11(a) shows $-\ln A$ vs $\delta(T)$ for $V_3Ga_{0.9}Sn_{0.1}$.
The value of S(T) at low T is proportional to

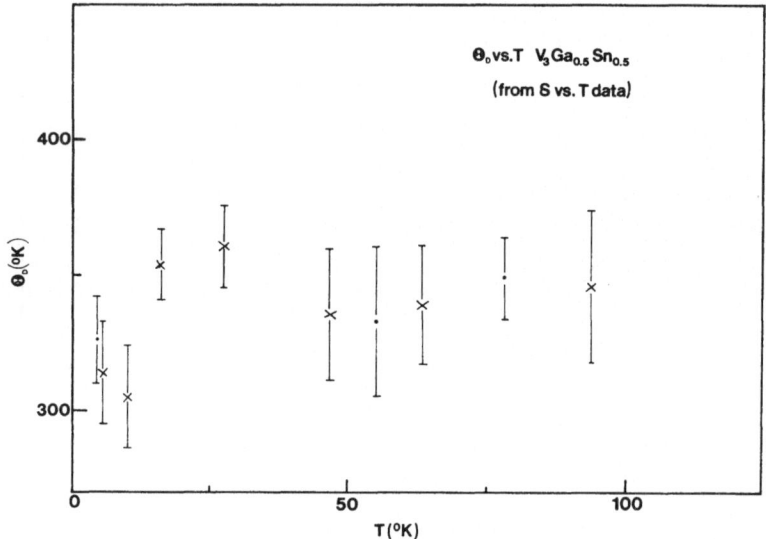

Fig. 10   Temperature dependence of the Debye temperature
from the thermal shift for $V_3Ga_{0.5}Sn_{0.5}$.

the phonon moments in the McMillan expression for λ,

$$S(0) \sim \frac{1}{<\omega>_0/<\omega^{-1}>_0} \quad .$$

The statistical scatter in the data near $T=0°K$ precludes
an experimental extrapolation to obtain $S(0)$.[6] For a similar
problem, Kitchens et al. have used the results of a linear
extrapolation from high temperature.[6] Here, however, an
analytic calculation of the dependence of $-\ln A$ on $\delta(T)$ is
made by combining the different Debye models for $<\omega^{-1}>$
($\sim <x^2> \sim -\ln A$) and $<\omega>$ ($\sim <v^2> \sim \delta(T)$).   The curve which
results from this analysis is the solid line in Fig. 10;
the slope of the curve at low temperature differs both from
that at high temperature and the expectation on the basis
of a single Debye temperature.   An average slope S_X at low
temperature derived from the numerical analysis is given in
Table I for all three concentrations (normalized to the
0.9 Ga result).   We emphasize that different Debye spectra

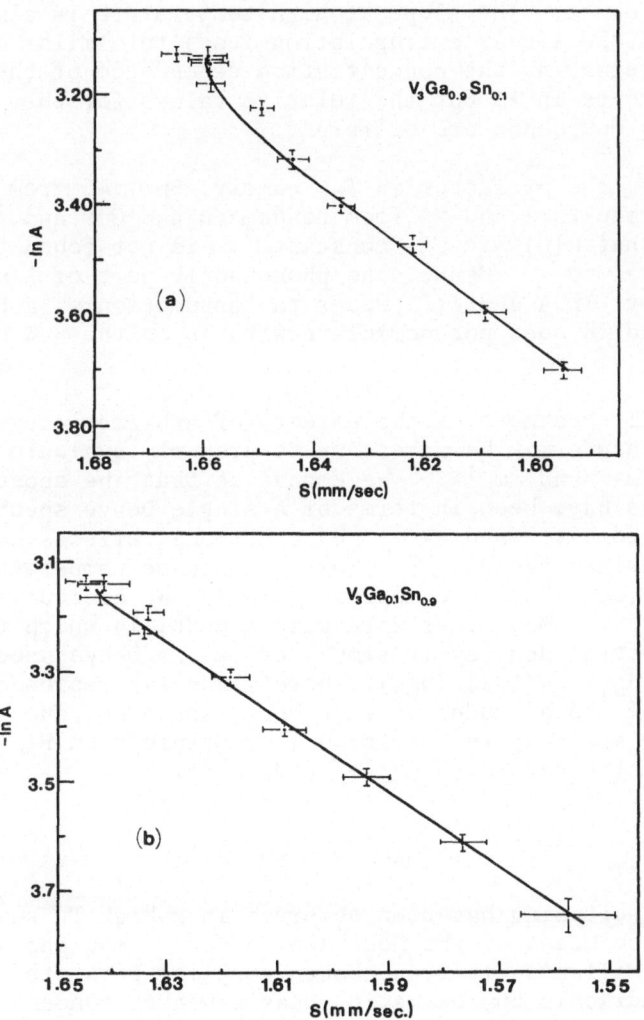

Fig. 11   (a) -ln A vs $\delta_{th}(T)$ for $V_3Ga_{0.9}Sn_{0.1}$.   (b) -ln A
vs $\delta_{th}(T)$ for $V_3Ga_{0.1}Sn_{0.9}$. The solid line is the curve
obtained with a Debye model for $<x^2>$ and $<v^2>$ and the
Debye temperatures given in Table I.

are necessary to reproduce the temperature dependence of $\delta(T)$ and $-\ln A$ and this fact is used in calculating their interdependence. The slope at high temperature is almost linear and the linear extrapolation leads to similar conclusions regarding the concentration dependence of the phonon moments in λ, but the relative values for the $V_3Ga_xSn_{1-x}$ compounds are different.

Taking the prefactor in T_c, namely, $\theta_D/1.45$ from specific heat results,[4] and μ^* from Bennemann and Garland,[16] and assuming that $N(0) <I^2>$ is constant,[7] S_x is not found to be proportional to λ. Hence, the phonon-only part of λ as measured by $-\ln A$ and $\delta(T)$ under the assumption of a quasiharmonic solid does not correlate with T_c in the β-W system, $V_3Ga_xSn_{1-x}$.

A full treatment of the effects of anharmonicity at low temperature may be necessary to properly evaluate the moments, $<\omega>_0$ and $<\omega^{-1}>_0$. We emphasize that the above considerations have been in terms of a single Debye spectrum, different for $<v^2>$ and $<x^2>$, and anharmonic effects have been taken into account by allowing θ_D to be temperature dependent (quasiharmonic approximation). We are currently reanalyzing the Mössbauer data with a model in which the phonon spectral density is simulated by two Debye spectra, $F(\omega) = aF(\theta_{D_1}) + (1-a)F(\theta_{D_2})$, where a and $1-a$ represent the fraction of total modes in each Debye spectrum, and concomitantly applying the additional constraints on $F(\omega)$ which arise from the sum rules on $<\omega>$ and $<\omega^{-1}>$.

CONCLUSIONS

Mode softening has been observed in a high T_c superconductor by means of the Mössbauer effect. The phonon-only part of McMillan's electron-phonon coupling parameter λ, in the quasiharmonic approximation, has a weaker concentration dependence than T_c.

REFERENCES

1. M. Weger and I.B. Goldberg in Solid State Physics, vol. 28, ed. H. Ehrenreich, F. Seitz and D. Turnbull, (Academic Press, New York, 1973) p. 2.

2.   J.S. Shier and R.D. Taylor, Solid State Comm. 5, 147
     (1967); Phys. Rev. 174, 346 (1968).
3.   L.J. Vieland, Phys. Rev. B 3, 1804 (1971).
4.   G. Knapp and H. Culbert, Bull. Am Phys. Soc. 19, 228 (1974).
5.   R.D. Taylor and P.P. Craig, Phys. Rev. 175, 782 (1968).
6.   T.A. Kitchens, P.P. Craig and R.D. Taylor in Mössbauer
     Effect Methodology, Vol. 5, ed. I. Gruverman, (Plenum
     Press, New York, 1969) p. 123.
7.   W.L. McMillan, Phys. Rev., 167, 331 (1968).
8.   J.M. Rowell and R.C. Dynes, in Phonons ed. M.A. Nusi-
     movici (Flammerion Press, Paris, 1971) p. 150.
9.   S. Ruby (Private communication).
10. W. B. Pearson, The Crystal Chemistry and Physics of
     Metals and Alloys, (Wiley-Interscience, New York, 1973).
11. H.S. Moller and R.L. Mössbauer, Phys. Letters 24A,
     416 (1967).
12. L.R. Testardi, Phys. Rev. B 5, 4342 (19).
13. F.Y. Fradin and D. Zamir, Phys. Rev. B 7, 4861 (1973).
14. L.R. Testardi and T.B. Batemen, Phys. Rev. 154, 402
     (1967).
15. L.R. Testardi et al., Phys. Rev. 154, 399 (1967).
16. K. Bennemann and J. Garland, AIP Conference Proceedings
     4, 103 (1972).

A MÖSSBAUER OBSERVATION OF ANISOTROPIC DIFFUSION NEAR THE GLASS TRANSITION OF A SMECTIC H LIQUID CRYSTAL*

R. E. DETJEN AND D. L. UHRICH

Department of Physics and Liquid Crystal Institute
Kent State University
Kent, Ohio   44242

We have observed the diffusion anisotropy of the iron bearing probe molecule, 1,1' - diacetylferrocene (DAF), near the glass-supercooled smectic H liquid crystalline transition in 4-n-hexoxybenzylidene-4'-n-propylaniline (HBPA). The sample which contained 0.2% by weight of DAF (enriched ~ 50% in Fe-57) in the HBPA, was oriented by cooling through the nematic phase to the smectic H phase in the presence of a 9000 gauss magnetic field. The resonance linewidth was then observed as a function of temperature for $\Theta = 0^\circ$ and $\Theta = 90^\circ$ orientations, where Θ is the angle between the preferred molecular direction as determined by the magnetic field and the gamma beam direction. Below 233°K the linewidth remained constant. From $233^\circ - 237^\circ$K diffusive broadening occurs such that at 237°K the rate of diffusion for $\Theta = 90^\circ$ was approximately four times greater than that for the $\Theta = 0^\circ$ orientation. The so-called 'free volume' model predicts that the onset of diffusion occurs at the supercooled liquid crystalline-glass transition temperature (T_g). As a result, we have assigned $T_g = 233^\circ$K for HBPA.

INTRODUCTION

The supercooled liquid-glass transition has been the object of much study via the Mossbauer technique over the past few years. In particular, measurements in aqueous solutions[1] and organic solvents such as propane[2] and methonal[3] have been reported. Polymers have also been subjects for this type of study.[4] The reasons for this activity are manyfold and include the fact that several of the Mossbauer

113

parameters, such as, recoil-free fraction (f), linewidth (Γ), and quadrupole splitting (ΔE_Q), have been found to change, sometimes dramatically, at the glass transition temperature (T_g). However, the above systems do not exhibit identical behavior at the glass transition. For example, for Fe^{+2} in $H_3PO_4 \cdot H_2O$ the linewidth, recoil-free fraction and the quadrupole splitting all undergo severe change at the glass transition. In contrast, for $Fe(ClO_4)_2 \cdot 6H_2O$ in the polymer, polyacrylonitrile, there is no change in the quadrupole splitting, or the linewidth through the glass transition and the recoil-free fraction has a slope discontinuity at T_g when plotted (as %) against the temperature.

Recently, researchers have become interested in the glassy states of several liquid crystalline phases and in the supercooled liquid crystal-glass transitions in these materials.[5,6] In this report, we have made use of the alignment properties of the liquid crystalline phases to observe anisotropic line-broadening and therefore, anisotropic diffusion of the impurity molecule 1,1' - diacetylferrocene (DAF) near the supercooled smectic H liquid crystal-glass transition in 4-n-hexoxybenzylidene-4'-n-propylaniline (HBPA) To our knowledge the only other Mossbauer observation of anisotropic diffusion was for Fe-57 in single crystals of graphite.[7] The solution used for study contained 0.2% by weight of DAF, (enriched to ~ 50% Fe-57) in HBPA. We have recently reported on this solution in a temperature study of the recoil-free fraction (f) from 100° - $230^\circ K$.[9] The smectic H glass was found to deviate from Debye-like behavior in the range 175° - $230^\circ K$. Below $175^\circ K$ the linear dependence of the $\ln f$ vs. temperature plot yielded a Debye temperature of $66^\circ K$.

Many of the properties of the liquid crystalline phase of HBPA have been reported elsewhere.[10,11] In particular, we have used its alignment properties to create ordered solute "monocrystals" of tin bearing molecules for the purpose of determining the sign of the nuclear quadrupole interaction (eQV_{zz}) and the molecular and lattice contributions (at $77^\circ K$) to the nuclear vibrational asymmetry. The fact that alignment initiated in the nematic phase by a magnetic field will persist to below T_g is the necessary requirement for the anisotropic diffusion measurements of this paper.

Briefly, the rather long HBPA molecules (~ 28 A) are

easily aligned in the nematic phase by the magnetic field.
The temperature of the system is then lowered through the
smectic A phase to the smectic H phase which readily super-
cools to room temperature. The parallel alignment effected
in the nematic phase is retained in the smectic A and smectic
H phases. The smectic H phase which is the most viscous of
these layered liquid crystalline structures is characterized
by a 2-dimensional hexagonal lattice and tilt of the molecular long axes from the planar normal. Furthermore, some x-
ray work suggests that there may be a 3-dimensional monoclinic lattice.[12] McMillan and Meyer[13] have proposed that
the tilt is caused by lack of complete rotation of the molecules about their long axes which allows the electric dipoles
on different molecules to interact. If complete rotation is
allowed, then the tilt angle goes to zero due to the averaging of the dipoles perpendicular to the long axes and the
result is the smectic B phase. As yet this theory has not
been completely verified but nuclear magnetic resonance (nmr)
work has established that there is in fact a partial freezing
out of the rotation of the molecules about their long axes in
the smectic H phase.[14]

The details of sample properties and the experimental
procedure used in obtaining the anisotropic diffusion measurements will be given in what follows, along with a discussion
of the results in terms of the liquid crystalline structure
and the glass transition.

EXPERIMENTAL

The liquid crystalline material HBPA has the following
phase transitions:

$$\begin{array}{ccccc}
29^\circ C & 66^\circ C & 68^\circ C & 85^\circ C \\
\text{Solid} \rightarrow \text{Smectic H} \updownarrow \text{Smectic A} \updownarrow \text{Nematic} \updownarrow \text{Isotropic}
\end{array}$$

The sample used for this study was prepared by dissolving
0.2% by weight of DAF in the isotropic phase of HBPA. The
solution integrity was checked by differential thermal analysis (dta), microscopic evaluation, and x-ray analysis and
has been reported in detail, previously.[8,9] The method of
alignment has also been reported previously and consists of
cooling the disc-shaped sample (~ 2mm thick) from the
nematic phase to room temperature in a 9000 gauss magnetic
field.[10,11] Here the sample remained in the supercooled
smectic H phase and the field could be removed without af-
fecting the alignment. It was then inserted in the Mossbauer

absorber dewar and cooled to the desired temperature for the
Mossbauer experiment. This study of anisotropic diffusion
required measurements for both parallel ($\Theta = 0°$) and perpendicular ($\Theta = 90°$) orientations. That is, the sample was prepared for the Mossbauer experiment so that the preferred molecular direction (as determined by the magnetic field) was
either parallel or perpendicular to the normal to the face
of the sample disc and thus either parallel ($0 = 0°$) or
perpendicular ($\Theta = 90°$) to the gamma direction. Unfortunately, due to the characteristic tilt angle of the smectic H
phase the planes could not be aligned. Since only the molecular preferred direction is determined by the field, all
values of the azimuthal angle of the planar normal were
equally probable. A better choice for these measurements
would have been a smectic B liquid crystal because in the
absence of a tilt angle the diffusion anisotropy would be
with respect to a complete set of aligned planes. Therefore,
diffusion within the layer could be compared to diffusion
from one layer to another. There are two reasons why HBPA
was chosen for these studies instead of a smectic B: first,
the experimental ease of handling the room temperature smectic
and second, the known integrity of the solution from the
previous recoil-free fraction study. However, since this
work was initiated, room temperature smectic B compounds have
been reported[15] and are presently being prepared for study.

The Mossbauer experiments were limited to three hours
(background ~ 165,000 counts) in the temperature range where
broadening occurred. For longer experimental run times the
sample would begin to crystallize and cause the Mossbauer
percent effect to increase with time. That the sample had
not crystallized was checked after each run by repeating the
run immediately. One could tell if crystallization had occurred by the rapid growth of unbroadened lines. In the
temperature range ($233° - 237°K$) in which diffusion broadening
occurred, as many as five three hour runs were added together
to provide adequate statistics for analysis. The added
spectra were fit with two lorentzian lines and the positions,
linewidths, and intensities determined from the computer fits.
To present better statistical representations of the data,
we have folded the spectra about the computer designated
centroid of the two line spectrum and fit the result with a
single lorentzian line. This procedure did not affect in any
way the observed broadening trends and besides providing a
better statistical display of the data, it was a simple way
of obtaining an average linewidth from the quadrupole split

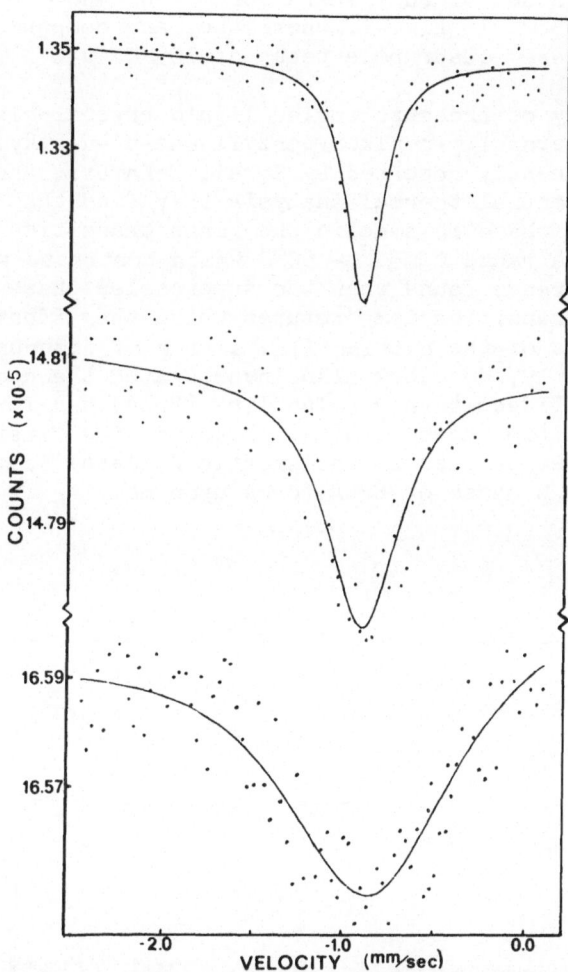

Fig. 1.   The Mossbauer transmission spectra for 0.2% (by
weight) of DAF in HBPA are shown (from top to bottom) for
T = 125°K, T = 237°K and Θ = 0°, and T = 237°K and Θ = 90°,
respectively.   Here Θ is the experimental angle between the
preferred molecular direction and the gamma beam from a
Co-57 in Cu source.   The upper spectrum is one member of the
quadrupole split doublet and is an example of the unbroadened
resonance.   The 237°K data exhibit broadening and here, for
display purposes, the doublet spectra have been folded about
the computer fit centroid.   The solid lines are the best fit
of lorentzians to the data.

spectra. These folded spectra for $\theta = 0°$ and $\theta = 90°$ at 237°K are shown in Fig. 1. where they are compared to one of the unbroadened quadrupole peaks at 125°K.

A study of the supercooled liquid crystal-glass transition in several n-p-alkoxybenzylidene-p'-n-alkylanilines has been recently reported by Sorai, Nakamura, and Seki.[6] From differential thermal analysis they find that if the supercooled phase is smectic the glass transition phenomena occur over a range of 30° - 60°C which contrasts with the 10° - 15°C range found when the supercooled phase is nematic. The glass transition temperatures which they identified for five members of the butylaniline series of compounds ranged from 158° - 197°K. They also investigated the compound 4-n-butoxybenzylidene-4'-n-ethylaniline (BBEA) and found no glass transition down to 88°K. BBEA was the first compound ever reported to possess the smectic H classification.[12] The smectic H phase of BBEA forms upon melting from the

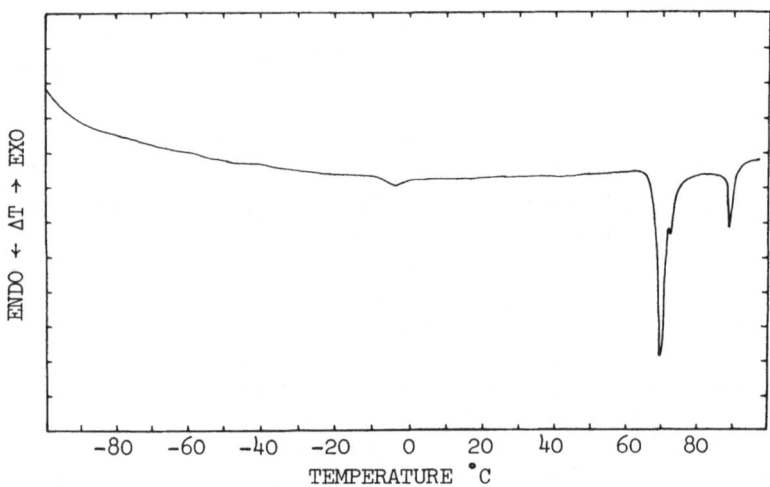

Fig. 2. The differential thermal analysis heating thermogram (15°C/min) for HBPA between -100° and +100°C. On heating the endothermic peaks correspond to the smectic H - smectic A, (~ 66°C), smectic A - nematic (~ 68°C), and nematic-isotropic (~ 85°C) transitions, respectively.

crystalline solid phase at 41°C. However, Sorai et. al. find
an extra lower temperature smectic phase (19° - 40°C).
Leaving the literature disagreement aside, the point here is
that differential thermal analysis of HBPA also did not exhibit a glass transition above 143°K. Fig. 2 shows the
differential thermogram (heating) illustrating the high
temperature phases and a small endothermic peak at -10°C
(263°K). This does not look like a glass transition which
is usually characterized by a step and does not correspond
to any crystallization which would be a large exothermic
peak. In addition, it is 30°K above the onset of the
Mossbauer line-broadening (see Fig. 3) and 88°K above the
temperature where the $\ln$ f deviates from the predictions of

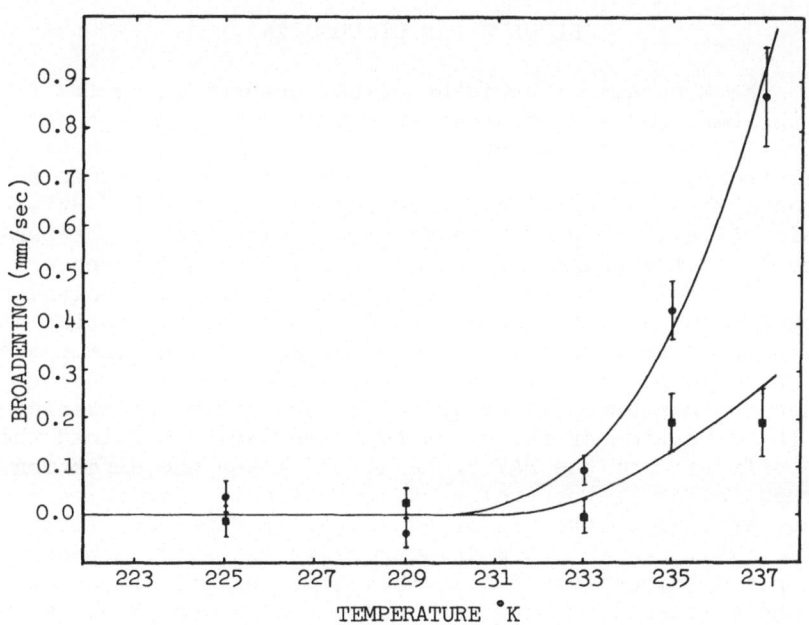

Fig. 3. A plot of line broadening vs. temperature for DAF
in HBPA. Both the $\Theta = 0°$ (■) and the $\Theta = 90°$ (●) orientations are displayed. Here Θ is the angle between the preferred molecular direction (as determined by the aligning
magnetic field) and the gamma direction.

the Debye model. This peak is reproducible and present in
both the 0.2% solution of DAF in HBPA and in pure HBPA. It
is probably associated with a relaxation of some molecular
motions in the supercooled lattice. Sorai et. al. suggest
that the long temperature range occupied by glass transition
phenomena in smectics is due to the fact that the relaxation
time of molecular motions cannot be described by a single
value. It is likely that in the HBPA the glass transition
phenomena are spread out over a large temperature range above
143°K and as a result the step in the dta thermogram which
is characteristic of the glass transition is too broad to be
observed. However, the Mossbauer effect which perceives molecular motions at the molecular level specifically determines
the onset of relaxation associated with two separate types of
molecular motion, namely, vibration and diffusion. Because
of the alignment properties of the liquid crystalline phase
both are anisotropic.

RESULTS AND DISCUSSION

     The Mossbauer linewidths of the quadrupole split lines
of the DAF remained constant at ~ 0.30 $\pm$ 0.03 mm/sec from
100° - 233°K, independent of orientation. Fig. 3 shows the
anisotropic broadening in the narrow temperature range be-
tween 233° - 237°K. The broadening for the $\Theta = 90^{\circ}$ orientation (i.e., the preferred molecular direction is perpen-
dicular to the gamma direction) is significantly greater than
for the $\Theta = 0^{\circ}$ orientation. In 1960, Singwi and Sjolander
showed that line broadening was due to diffusive motion of
the probe molecules and they also suggested the measurement
of anisotropic diffusion in graphite single crystals.[16]
Since the measured diffusive motion due to the broadening is
in the direction of the gamma beam, we have determined that
the diffusion of the DAF probe at 237°K for the direction
perpendicular to the preferred direction is four times greater
than parallel to the preferred molecular direction (i.e.,
$D_{\perp}/D_{\parallel}$ ~ 4). Again, this anisotropy is not with respect to
the planar directions due to the characteristic tilt angle
of the smectic H. There is Mossbauer evidence which, when
coupled with the x-ray value of 34° for the tilt angle in
BBEA[12], yields the tilt angle in HBPA to be ~ 30°.[11] This
means that the projection (p) of the unit planar normal along
the $\Theta = 0^{\circ}$ direction (p = 0.87) is greater than its projection ($0.5 \geq p \geq 0$) along the $\Theta = 90^{\circ}$ direction. As a result,
diffusion for the $\Theta = 0^{\circ}$ direction includes a larger component of the diffusion normal to the layers than does the

diffusion for $\Theta = 90^\circ$. Since the broadening for the $\Theta = 0^\circ$ orientation (and therefore most representative of the diffusion normal to the layers) is less than the broadening for $\Theta = 90^\circ$ (and therefore most representative of diffusion within the layers) the data of this paper agree with known observations of diffusion anisotropy in smectic liquid crystals. In particular, an nmr study of the diffusion of tetramethyl-silane (TMS) in the smectic A and B phases (not supercooled) of 4-n-butoxybenzylidene-4'-n-octylaniline demonstrated that the diffusion of the impurity within the layer was significatly greater than diffusion normal to the layers.[17] The DAF probe of this study is similar to the TMS nmr probe in that both are much smaller than the host liquid crystal molecules.

The Mossbauer effect is usually unable to distinguish whether diffusion occurs by continuous diffusion or by jump diffusion. The first corresponds to diffusion proceeding via a large number of small jumps such that no single jump destroys the Mossbauer resonance. For this case the line broadening is given by:

$$\Delta\Gamma = 2\hbar k^2 D \tag{1}$$

where k is the gamma ray wave vector and D is the diffusion constant. If on the other hand the diffusion proceeds via large uncorrelated jumps (such that $< x^2 >$ for those molecules which had jumped was large and their corresponding recoil-free fraction was effectively zero) then the line broadening is given by:

$$\Delta\Gamma = 2\hbar/t_j \tag{2}$$

where t_j is the average time between jumps for the probe molecules. The data of Fig. 3 have been reduced to correspond to each model and the results are listed in Table I. Through the narrow temperature range where broadening occurs the isomer shift does not change to within experimental error. Also the quadrupole splitting appears to remain constant ($\sim$ 2.15 mm/sec) except for both runs at 237°K. Here, the computer fit shows a decrease of ΔE_Q of approximately 10%. However, more evidence is required to establish

Table I

Θ	$T({}^{\circ}K)$	Γ(mm/sec)	$\Delta\Gamma$ (mm/sec)	$D(cm^2/sec)$ $\times 10^{11}$	t_j(sec) $\times 10^7$
0	233°	0.29	-0.01	0.0	∞
0	235°	0.50	0.20	1.36	1.35
0	237°	0.48	0.18	1.25	1.47
90	233°	0.39	0.09	0.61	3.01
90	235°	0.74	0.44	3.01	0.62
90	237°	1.16	0.86	5.83	0.32

if this decrease is indeed real.

There are two theories of the glass transition which lead to the result that translational diffusion is allowed only above T_g and obeys a modified form of the Arrhenius equation. [18,19] The first is the so-called free volume approach which assumes that the diffusive jumps occur when there is a redistribution of the surrounding molecules such that the local volume increase can accommodate the diffusing molecules. This local volume increase is facilitated by a redistribution of the 'free volume'. The second theory which was proposed by Adam and Gibbs preserves the notion that the free volume vanishes at T_g.[18] Furthermore, it is less restrictive and more generally applicable to complex systems. The main result is that the macroscopic configurational entropy of the glass forming liquid (liquid crystal) disappears at the glass transition. This result is obtained by assuming only that the cooperatively rearranging groups of molecules required for translational diffusion do not interact. Then the temperature dependence of the minimun size of these rearranging groups can be expressed in terms of the configurational entropy. If indeed the glass transition is governed by the configurational entropy then it will be only weakly dependent on the vibrational and rotational degrees of freedom. This was the result of Reich and Michaeli[4] who found a rapid increase in f with decreasing temperature below the glass transition for Fe-57 in the polymer, polyacrylonitrile. They ascribed this increase to the 'freezing in' of the rotational degrees of freedom in the glassy state. In this case, however, no broadening was observed through the glass transition.

The glass transition phenomena in smectic glasses are known to extend over a rather large temperature range.[6] In fact the dta data of this paper (see Fig. 2) show that the smectic H phase of HBPA does not exhibit the step characteristic of the glass transition. However, the Mossbauer data distinguish both the translational diffusion and the vibrational motion of the DAF probe on the molecular level. We have used the onset of diffusion as the determining factor in our assignment of T_g = 233°K in HBPA. This is in accordance with both of the above mentioned theories for the glass transition. To be noted here is that the DAF probe is approximately one-fourth the length of the HBPA molecule. Therefore, the size of the cooperatively rearranging group of molecules which is required for diffusion is less for DAF than for HBPA. As a result T_g = 233°K may be a low estimate of the true glass transition of the smectic H phase of HBPA.

Previously, we have reported the temperature dependence of the ℓn f for DAF in HBPA.[8,9] On heating non-Debye behavior started at ~ 175°K. Furthermore the ℓn f vs. T plot shows f rapidly tending to zero in the temperature range of the line broadening. On cooling we can attribute this non-Debye behavior to the 'freezing in' of the vibrational degrees of freedom associated with the rather long end chains of the HBPA molecule.

CONCLUSION

We have observed, via the Mossbauer effect, the anisotropic diffusion of the DAF probe molecule in the 'frozen in' smectic H structure of HBPA. The onset of diffusion at 233°K has been identified as the glass transition in HBPA in accordance with theory.[18,19] Furthermore, the anisotropy is in qualitative agreement with other experiments in the normal smectic phases which show that translation normal to the layers is more hindered than that within the layers. There are, however, two weaknesses to the data intrepretation presented here. The first is the lack of an independent measurement of the glass transition by a technique other than the Mossbauer effect and the second is the inability to quantitatively compare the inter-layer diffusion with the intra-layer diffusion. To overcome both difficulties we are preparing to study a smectic B liquid crystal for which the dta shows a measureable step at T_g. Thus, the assignment of T_g will be unambiguous and the absence of the molecular tilt in the smectic B structure will allow direct comparison of inter- and intra-layer diffusion.

ACKNOWLEDGEMENT

We gratefully acknowledge W. LaPrice for the dta thermogram shown in Fig. 2.

REFERENCES

*   This research has been supported in part by the NSF
       (Grant #GH-34164X) and the AFOSR (Contract F44620-69-
       C-0021).

1.   S. L. Ruby and I. Pelah, "Crystals, Supercooled Liquids
       and Glasses in Frozen Aqueous Solutions," in Mossbauer
       Effect Methodology, Vol. 6, I. J. Gruverman, Ed., Plenum
       Press, New York (1971); S. L. Ruby, "Mossbauer Studies
       of Aqueous Liquids and Glasses," in Prospectives in
       Mossbauer Spectroscopy, S. G. Cohen and M. Pasternak,
       Eds., Plenum Press, New York (1973); A. Abras and J. G.
       Mullen, Phys. Rev. 6A, 2343 (1972).

2.   J. H. Jenson, Phys. Kondens. Materie 13, 273 (1971).

3.   A. Simonpoulus, H. Wickman, A. Kostikas, and D. Petride
       Chem. Phys. Letters 7, 615 (1970).

4.   S. Reich and I. Michaeli, J. Chem. Phys. 56, 2350 (1972

5.   S.E.B. Petrie, H. K. Bucher, R. T. Klingbiel, and R. I.
       Rose, Eastman Organic Chemical Bulletin 45, No. 2 (1973

6.   M. Sorai, T. Nakamura, and S. Seki, Proceedings of the
       International Conference on Liquid Crystals, Banglalore
       India, 1973 (to be published).

7.   H. Pollak, M. de Coster, and S. Amelinckx, in Pro-
       ceedings of the Second International Conference on the
       Mossbauer Effect, D. M. Compton and A. H. Schoen, Eds.,
       John Wiley & Sons, New York (1962) p 112.

8.   R. E. Detjen, D. L. Uhrich, and C. F. Sheley, Phys.
       Letters 24A, 522 (1973).

9.   D. L. Uhrich, R. E. Detjen, and J. M. Wilson, "The Use
       of Liquid Crystals in Mossbauer Studies and the Use of

the Mossbauer Effect in Liquid Crystal Studies," in
Mossbauer Effect Methodology, Vol. 8, I. J. Gruverman
and C. W. Seidel Eds., Plenum Press, New York (1973).

10.   D. L. Uhrich, Y. Y. Hsu, D. L. Fishel, and J. M. Wilson,
      Mol. Cryst. and Liq. Cryst. 20, 349 (1973).

11.   D. L. Uhrich, J. Stroh, R. D'Sidocky and D. L. Fishel,
      Chem. Phys. Letters (to be published).

12.   A. DeVries and D. L. Fishel, Mol. Cryst. and Liq.
      Cryst., 16, 311 (1972).

13.   W. L. McMillan and R. Meyer, Phys. Rev. (to be
      published).

14.   J. W. Doane (private communication).

15.   Y. Y. Hsu, Ph.D. Dissertation, Kent State University
      (1972).

16.   K. S. Singwi and A. Sjolander, Phys. Rev. 120, 1093
      (1960).

17.   J. A. Murphy, J. W. Doane, Y. Y. Hsu, and D. L. Fishel,
      Mol. Cryst. and Liq. Cryst. 22, p 133 (1973).

18.   G. Adam and J. H. Gibbs, J. Chem. Phys. 43, 139 (1965).

19.   C. A. Angell, J. Phys. Chem. 70, 2793 (1966).

MÖSSBAUER EFFECT STUDIES OF ELECTRONIC RELAXATION IN FERRIC COMPOUNDS

S. Mørup

Laboratory of Applied Physics II
Technical University of Denmark
DK 2800 Lyngby, Denmark

The Mössbauer spectra of several paramagnetic ferric compounds consist of a relaxation broadened singlet or quadrupole doublet. The examination of the field dependence and the temperature dependence of these spectra makes it possible to study several effects.

Ferric ions with cubic surroundings have a small crystal field splitting. It appears that the Mössbauer spectrum in this case consists of a single, broad Lorentzian line. This suggests that the relaxation is isotropic. A magnetic field, which is large compared to the magnetic dipole field and the exchange field, gives rise to longitudinal relaxation and therefore changes the spectrum drastically. If the ferric ions have non-cubic surroundings an asymmetric quadrupole split Mössbauer spectrum can be found. In some cases an examination of the line shape and its dependence on the temperature and on the applied field enables one to estimate the size of the crystal field splitting.

An anomalous field dependence of the spin-spin re-

laxation time has been found in $Fe(NO_3)_3 \cdot 9H_2O$ and in frozen
solutions of ferric salts. In some cases Mössbauer spectroscopy is found to be very suitable for the study of the
temperature dependence of spin-lattice relaxation.

1. INTRODUCTION

Mössbauer spectra of paramagnetic compounds depend
strongly on the electronic relaxation time, τ. When the e-
lectronic relaxation is slow compared to the nuclear Larmor
period, τ_L, magnetically split spectra can be found. When
$\tau \approx \tau_L$ spectra with broadened lines are found, and for
$\tau < \tau_L$ the magnetic splitting collapses, but a broadening
of the Mössbauer lines, due to the relaxation, is often observable for values of τ which are smaller than τ_L by a few
orders of magnitude.

In ferric compounds $\tau_L \approx 10^{-8}$ sec. The Mössbauer spectra often consist of broad singlets or doublets, i.e.
10^{-11} sec $\lesssim \tau \lesssim 10^{-9}$ sec. These broadened Mössbauer lines
are in some cases purely Lorentzian, in other cases they
are made up by broad and narrow components. The line width
and the line shape often depend on the temperature and the
applied magnetic field.

The aim of this work is to study the line shape and its
dependence on temperature and applied fields in order to obtain information about the terms of the Hamiltonian and the
relaxation processes.

2. THE SPIN HAMILTONIAN

Ferric ions in a weak ligand field are in the high spin state ($S = 5/2$ and $L = 0$). The spin Hamiltonian may be expressed as a sum of operators:

$$H_1 = H_{cf} + H_Z + H_{dd} + H_{ex} + H_{hf};$$

The terms represent the crystal field interaction, the Zeeman interaction due to external fields, the magnetic dipole-dipole and exchange interactions with neighbouring ions, and the hyperfine interaction.

The crystal field Hamiltonian is conventionally written as a sum of an axial, a rhombohedral and a cubic term:

$$H_{cf} = H_{ax} + H_{rh} + H_{cub};$$

where

$$H_{ax} = D\{S_z^2 - 1/3 \; S(S+1)\}$$

and

$$H_{rh} = E\{S_x^2 - S_y^2\}$$

H_{cub} contains terms of fourth order in S, but H_{cub} is often negligible compared to H_{ax} and H_{rh}. Furthermore, $H_{hf} = H_m + H_Q$ where H_m and H_Q describe the magnetic and the quadrupole interactions of the nucleus with its surroundings.

In the paramagnetic compounds discussed here, $H_{dd} + H_{ex}$ may often be treated as a perturbation which gives rise to spin-spin relaxation. A second type of relaxation is spin-lattice relaxation which is caused by the interaction between phonons and the electronic spins.

In paramagnetic compounds any of the terms of H_1 may be important. Therefore the Mössbauer spectra may display many different effects.

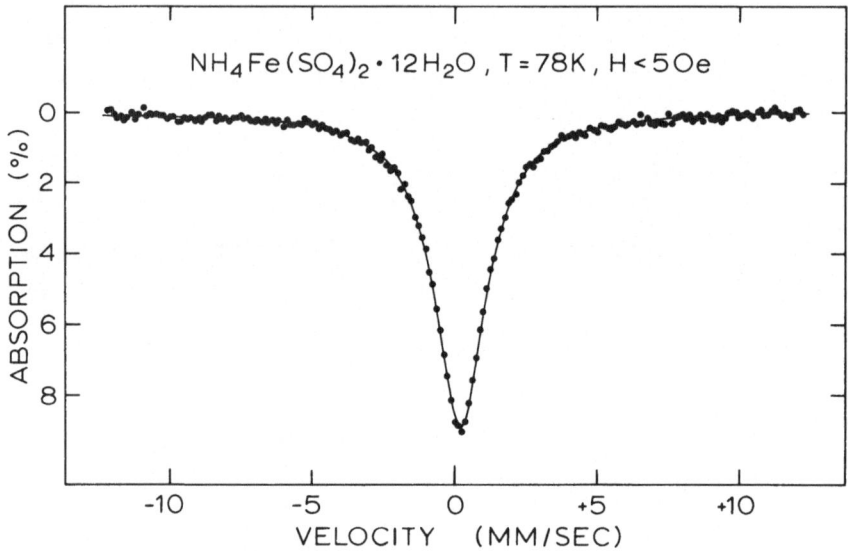

Figure 1. Mössbauer spectrum of ferric alum at H < 5 Oe.

In the following it will be discussed how the dominance
of specific terms of the Hamiltonian is reflected in the
Mössbauer spectra, at temperatures and external fields
where the magnetization is negligible.

3. $H_{dd} + H_{ex}$ ARE DOMINATING, $H_Q = 0$

The crystal field interaction of ferric ions in cubic
surroundings is often small. Then, in the absence of a
large external field field, $H_{dd} + H_{ex}$ may be large compared
to other terms of H_1. Two cubic compounds which exhibit relaxation broadened Mössbauer lines have been examined,
namely ferric alum [1-2] and $(NH_4)_3FeF_6$ [3]. A characteristic feature of these compounds is that their Mössbauer

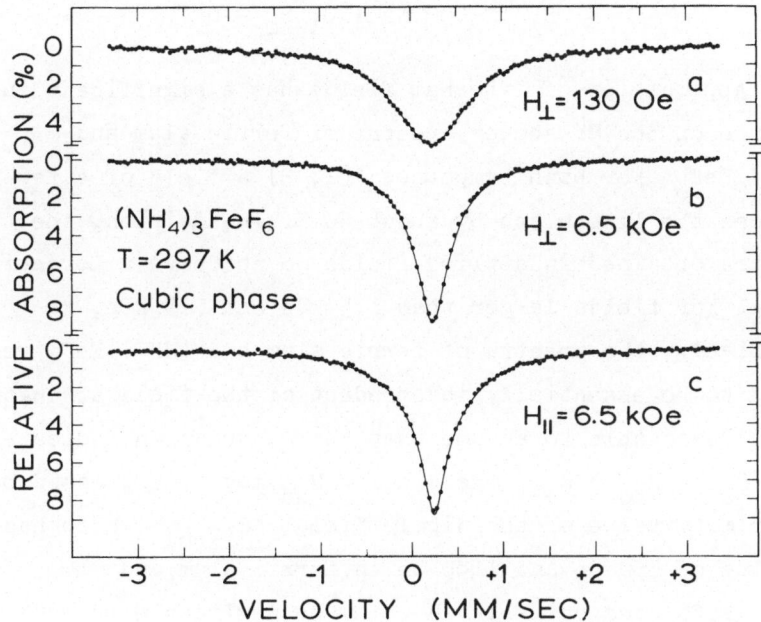

Figure 2. Mössbauer spectra of $(NH_4)_3Fe_6$ at various
applied fields [3].

spectra consist of a very broad, but purely Lorentzian line.
The spectra are shown in Fig. 1 and Fig. 2 (a) together
with Lorentzian fits.

When H_{cf} and H_Z are negligible the fluctuating hyperfine field, acting at the nucleus, has no preferred direction. Hence, the relaxation may be isotropic. Bradford
and Marshall [4] and Afanes'ev and Gorobchenko [5] have
computed the shape of the Mössbauer spectrum in the case
of isotropic relaxation. For $\tau < 10^{-9}$ sec the concurrent
result of these authors is that the spectrum is purely
Lorentzian. This is in excellent agreement with the measurements.

4. H_Z IS DOMINATING, $H_Q = 0$

Application of external fields has a significant influence on the Mössbauer spectra of ferric alum and
$(NH_4)_3FeF_6$. For both compounds [1, 3] a field of a few kOe
reduces the line width by about 50 %. In addition, the
spectra obtained in applied fields do not have a Lorentzian
shape. For fields larger than 1.5 kOe and 3.5 kOe, respectively, the spectra of ferric alum and $(NH_4)_3FeF_6$ were
found to be essentially independent of the field strength.
It is reasonable to assume that this saturation indicates
that $H_Z \gg H_{dd} + H_{ex}$. The size of H_{dd} may be expressed by
the r.m.s. value of the dipole field, $<H_i>_{rms}$, which has
a value of 450 Oe and 1180 Oe in ferric alum and
$(NH_4)_3FeF_6$, respectively. H_{ex} may be negligible in both
compounds as the distance between the ferric ions exceeds
several Angstroms.

When H_Z dominates, the eigenstates of the ferric ions
are essentially the eigenstates of S_Z. The electronic relaxation now causes a fluctuation of the hyperfine field
along the z-direction (longitudinal relaxation). Theoretical spectra for longitudinal relaxation have been computed by several authors [6-10]. When $\tau < \tau_L$ the spectrum
generally consists of three Lorentzian lines with a line
width given by [2]:

$$\Gamma = \Gamma_0 + \nu(m_e, m_g)^2 < h(t)^2>_{av} \tau$$

Γ_0 is the thin absorber line width in the absence of relaxation effects, $\nu(m_e, m_g)$ is the difference between the
z components of the nuclear magnetic moments of the excited

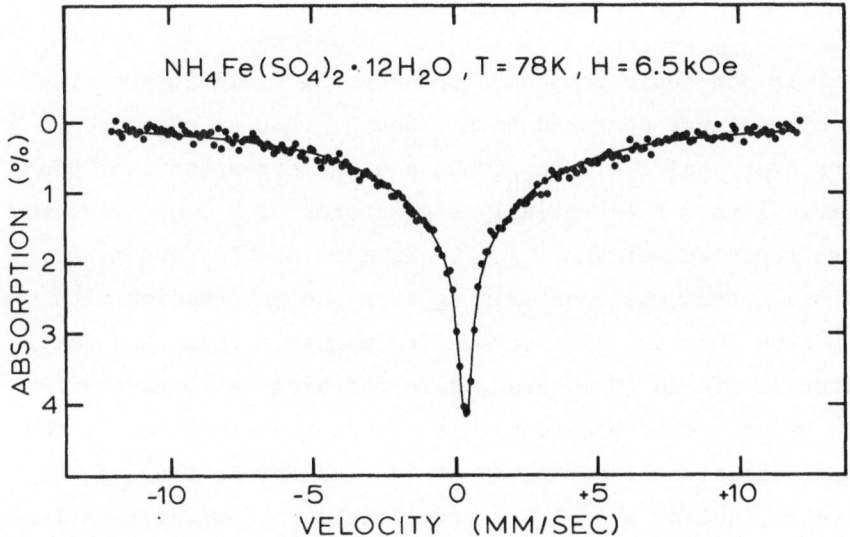

Figure 3. Mössbauer spectrum of ferric alum
           at H = 6.5 kOe.

state and the ground state, and h(t) is the time dependent
hyperfine field. The area ratios of the three components
are 3:4:1 when the field is perpendicular to the gamma ray
direction and 3:0:1 when the field is parallel to the gamma ray direction.

In Fig. 2 (b) and (c) are shown the spectra of
$(NH_4)_3FeF_6$ in applied fields of 6.5 kOe perpendicular and
parallel to the gamma ray direction, respectively, and in
Fig. 3 is shown a spectrum of ferric alum with the field
perpendicular to the gamma ray direction. The solid lines
represent the best fits based on theory.

5. H_{cf} IS DOMINATING

In non-cubic compounds the crystal field interaction is often large compared to H_{dd} and H_{ex}. In an axially symmetric crystal field, $H_{rh} = 0$, and the eigenstates of the ferric ions are essentially eigenstates of S_z. However, the wave functions of the $| \pm 1/2>$ Kramers doublet are mixed by small perturbations arising from the interaction with the iron nucleus, the surrounding magnetic ions, and small external fields. This results in off-diagonal elements in the hyperfine interaction. Therefore, the simple model for longitudinal relaxation cannot be used for fitting of the spectra, unless a small external field is applied parallel to the symmetry axis of the crystal field.

In $FeCl_3 \cdot 6H_2O$ the crystal field is nearly axially symmetric [11] and the Mössbauer spectrum consists of an asymmetric quadrupole splitting [12-14]. We have found that theoretical spectra for pure longitudinal relaxation have a more pronounced asymmetry than that found in the spectrum of $FeCl_3 \cdot 6H_2O$. The discrepancy is probably due to the off-diagonal elements of the hyperfine interaction.

In some cases the Mössbauer spectra are strongly temperature dependent due to the temperature dependence of the relative population of the Kramers doublets. For $D > 0$ the $| \pm 1/2>$ doublet has the lowest energy. As the spin-spin relaxation of this doublet is very fast, a decrease in the spin-spin relaxation time is expected when the temperature is decreased. This effect has been found in $FeCl_3 \cdot 6H_2O$ [12-14] and in hemin [15].

For D < 0 only the | ± 5/2> Kramers doublet is occu-
pied at low temperatures. This gives rise to a very long
spin-spin relaxation time. This effect has been found in
some iron dithiocarbamates [16]. A study of the temperature
dependence of the spin-spin relaxation time makes it possible to estimate the size of D.

It has been shown [9] that if the hyperfine field fluctuates parallel to the principal axis of an axially symmetric electric field gradient, the $\pm 3/2 \rightarrow \pm 1/2$ nuclear
transitions have broader lines than the $\pm 1/2 \rightarrow \pm 1/2$ and
the $\pm 1/2 \rightarrow \mp 1/2$ transitions. However, the opposite asymmetry is found when the hyperfine field fluctuates perpendicularly to the principal axis of the electric field gradient.

The direction of the hyperfine field can be changed by
application of external fields. Especially the wave-functions
of the | ± 1/2> Kramers doublet are modified even by small
fields, but if $H_Z \gtrsim H_{cf}$ also the | ± 3/2> and | ± 5/2> states
are changed. For $H_Z \gg H_{cf}$ the electronic wave functions are
essentially the eigenfunctions of S_Z. Therefore, when an external field is applied perpendicularly to the symmetry axis
of a single crystal absorber of a compound with axial symmetry, the asymmetry of the Mössbauer spectrum may be reduced and for large fields, the opposite asymmetry may be
found. For $H_Z \gg H_{cf}$ the spectrum is independent of the field
strength. In a polycrystalline absorber the situation is
more complicated, but for $FeCl_3 \cdot 6H_2O$, the net result of the
application of a magnetic field is that the broader line
becomes narrower, and the narrower line is essentially unaltered. This is shown in Fig. 4. We have found a similar,

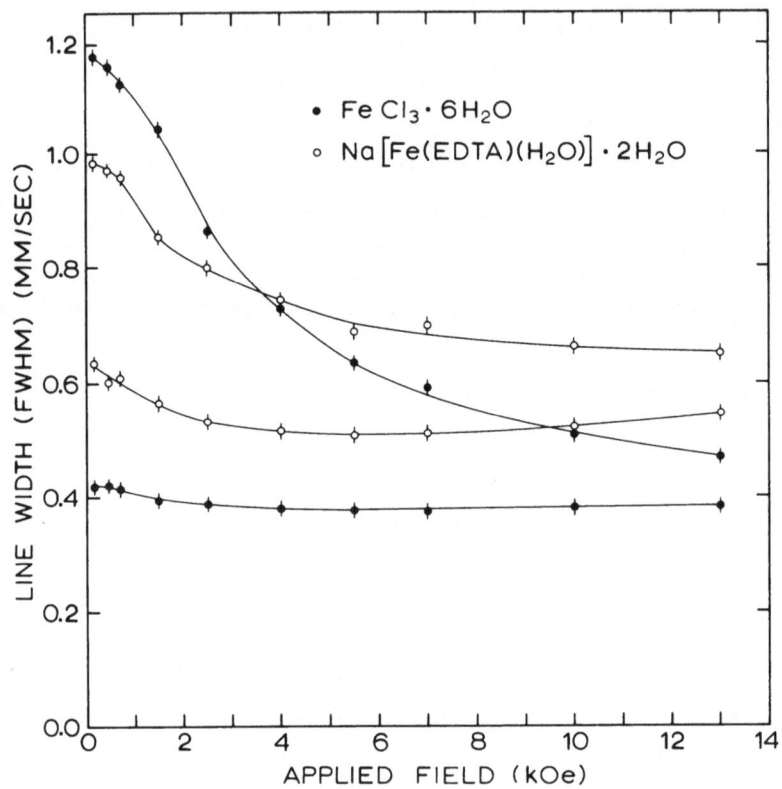

Figure 4. Mössbauer line widths of $FeCl_3 \cdot 6H_2O$ and
Na[Fe(EDTA)(H_2O)]$\cdot 2H_2O$ at 78 K as a function
of the applied field.

but less pronounced narrowing of the broad line of hemin.

In ferric EDTA complexes the ferric ions are seven co-
ordinate [17] and therefore the symmetry is not axial, i.e.
$H_{rh} \neq 0$, and the electric field gradient is not axially symmetric. The zero field Mössbauer spectrum of
Na[Fe(EDTA)(H_2O)]$\cdot 2H_2O$ consists of an asymmetric quadrupole
splitting, but the asymmetry is less pronounced than in

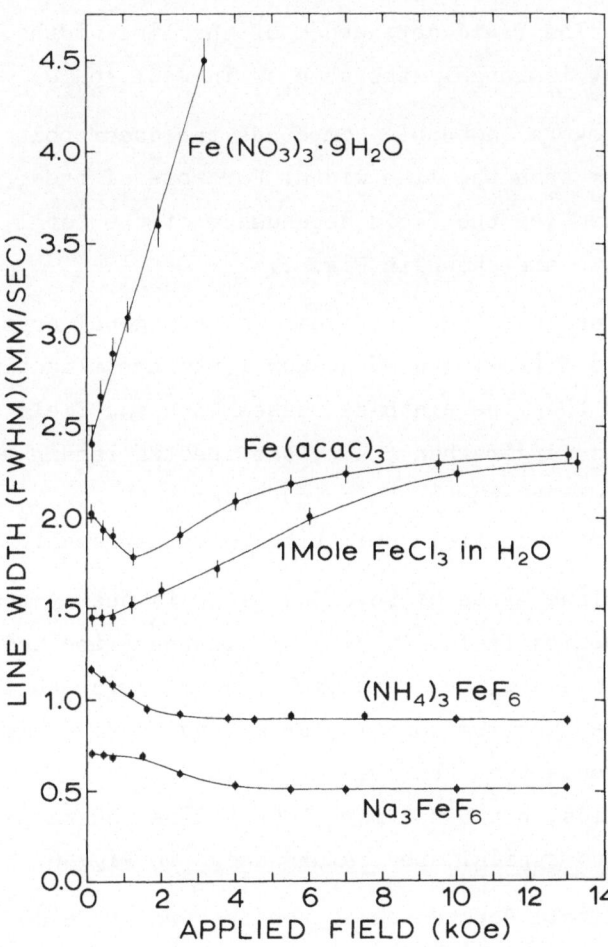

Figure 5. Mössbauer line widths (FWHM) of some ferric
          compounds as a function of the applied field.

$FeCl_3 \cdot 6H_2O$, i.e. the departure from theoretical spectra for longitudinal relaxation along the axis of an axial symmetric field gradient is larger than for $FeCl_3 \cdot 6H_2O$, as could be expected. The field dependence of the line widths, shown in Fig. 4, is roughly the same as in $FeCl_3 \cdot 6H_2O$.

In several noncubic compounds the quadrupole splitting is smaller than the line width. For some of these compounds we have studied the field dependence of the total line width. The results are shown in Fig. 5.

Na_3FeF_6 and $(NH_4)_3FeF_6$ have a tetragonal crystal structure at 78 K [18]. The Mössbauer lines are slightly asymmetric, and the line width decreases when the field is applied. For fields larger than 5 kOe, the spectra remain unaffected by the field strength. This indicates that now $H_z \gg H_{cf}$, i.e. the crystal field splitting is rather small.

The line width of $Fe(ClO_4)_3 \cdot 9H_2O$ is nearly independent of the applied field. At small fields the line nearly has a Lorentzian shape, but is slightly asymmetric. At large fields the spectrum contains broad and narrow components. This suggests that the relaxation is nearly isotropic at small fields, but has a more longitudinal character at large fields. Two typical spectra are shown in Fig. 6.

The field dependence of the spectra of $Fe(NO_3)_3 \cdot 9H_2O$, $Fe(acac)_3$ and frozen aqueous solutions of $FeCl_3$ is discussed in section 7.

6. INFLUENCE OF SMALL EXTERNAL FIELDS

In compounds with a small magnetic dipole-dipole interaction(i.e. a large interatomic distance between the

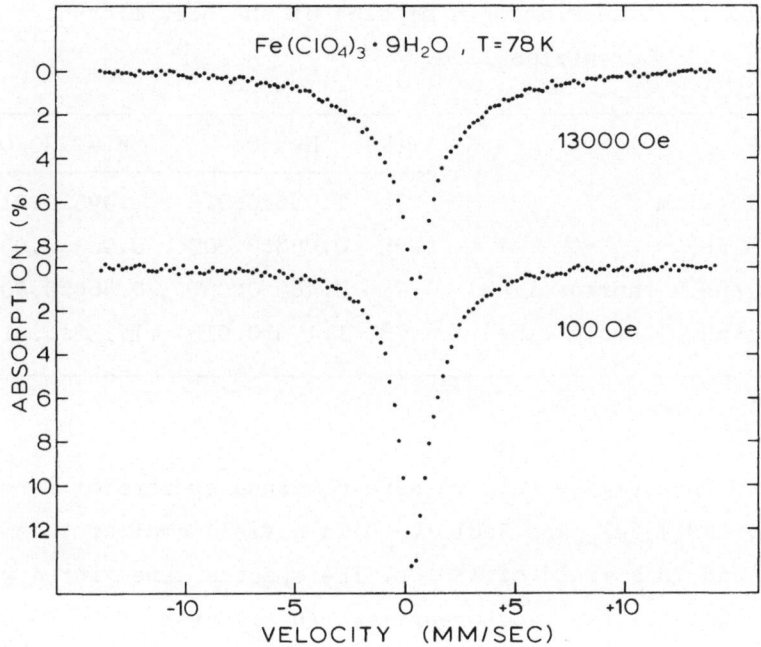

Figure 6. Mössbauer spectra of $Fe(ClO_4)_3 \cdot 9H_2O$ at 78 K.

ferric ions), magnetically split spectra can be found. Mössbauer spectra of this kind are greatly modified if the de-
generacy of the $|\pm 1/2\rangle$ Kramers doublet is lifted by ap-
plication of an external field of the order of 100 Oe [19-
20]

In the compounds discussed here the dipole-dipole field
$\langle H_i \rangle_{rms}$ is larger than 100 Oe. However, as H_i fluctuates
rapidly both in size and in direction, a small external
field might still have an influence on the spectra. In or-

Table 1. Line width (in mm/sec) of the best fitted
          Lorentzian line.

	T(K)	H<5 Oe	H = 130 Oe
Ferric alum	78	1.986±0.015	1.995±0.015
$(NH_4)_3FeF_6$	295	0.958±0.007	0.963±0.007
$FeCl_3 \cdot 6H_2O$ (narrow line)	78	0.485±0.007	0.464±0.007
$FeCl_3 \cdot 6H_2O$ (broad line)	78	1.155±0.015	1.132±0.015

der to investigate this we have obtained spectra of ferric
alum, $(NH_4)_3FeF_6$ and $FeCl_3 \cdot 6H_2O$ in a field smaller than
5 Oe and in a field of 130 Oe. The spectra were fitted with
Lorentzian lines (two Lorentzians in the case of
$FeCl_3 \cdot 6H_2O$). The best fits of all the spectra were within
the statistical uncertainty. The estimated line widths of
the spectra are given in Table 1. No significant influence
of the 130 Oe field was observed.

7. SPIN-SPIN RELAXATION

For most of the compounds which we have examined the
application of a magnetic field results in narrower lines.
However, there are a few exceptions to this rule. The line
width of a polycrystalline absorber of $Fe(NO_3)_3 \cdot 9H_2O$ at
78 K increases considerably when an external field is applied. This is shown in Fig. 5. For fields larger than
10 kOe side lines appear in the spectrum at about
±9 mm/sec [21]. This indicates that the relaxation time has
increased. As the spectra are independent of temperature

below 105 K, it can be concluded that spin-lattice relaxation is too slow to have any influence on the spectra.
Therefore, the change in the spectra must be due to a
field dependence of the spin-spin relaxation time.

In a compound where the ionic states are non-degene-
rate the most probable spin-spin relaxation processes consist in simultaneous transitions among identical eigenstates of two neighbouring ions. These processes (flip-
flop processes) nearly conserve the total energy.

In $Fe(NO_3)_3 \cdot 9H_2O$ the ferric ions are connected in
pairs by a screw diad axis and have identical energy levels.
However, when a field is applied, the energy splitting
strongly depends on the angle between the crystal field
axes and the magnetic field direction. Therefore, when the
field is applied, the two ions will generally have different
energy levels. This means that the flip-flop processes must
involve a change in the total energy, and therefore they
are not probable. Consequently, the relaxation time will
increase. In the special cases where the magnetic field is
parallel or perpendicular to the screw diad axis, the two
ions will have identical energy levels irrespective of the
size of the applied field. Therefore the relaxation time
should be shorter for these orientations than for others.
In fact, this has been demonstrated by measurements on
single crystals [22] in a magnetic field of 13 kOe. Some
spectra obtained for various orientations of the crystal
are shown in Fig. 7. θ indicates the angle between the screw
diad axis and the applied field. Obviously, for $\theta = 0^\circ$ and
for $\theta = 90^\circ$ the relaxation time is shorter than for other
values of θ.

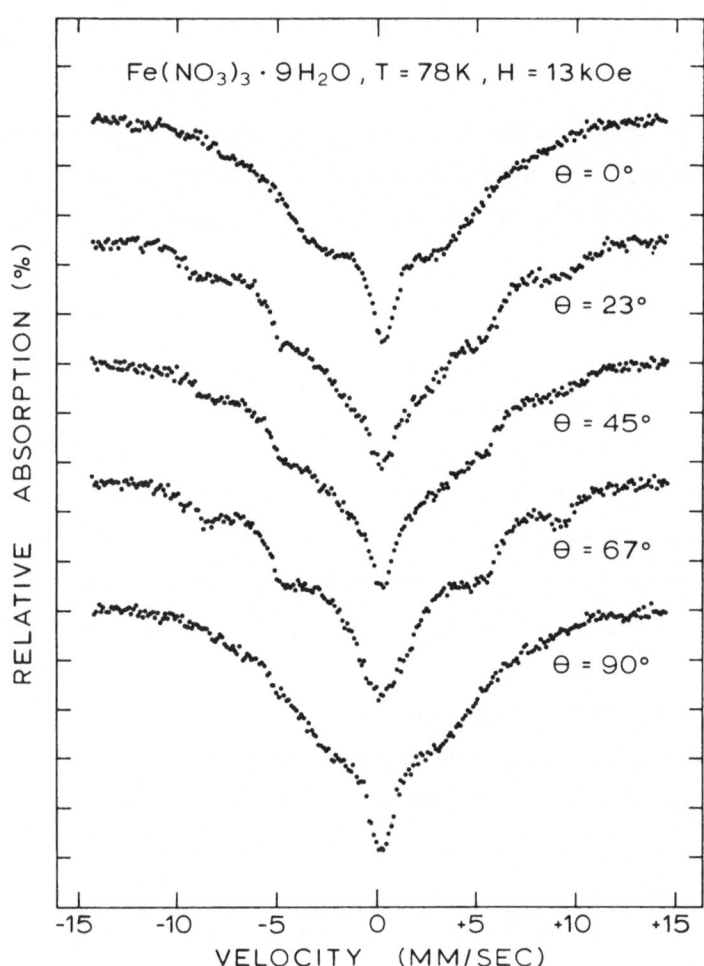

Figure 7. Mössbauer spectra of a single crystal absorber
of $Fe(NO_3)_3 \cdot 9H_2O$. θ is the angle between the
screw diad axis and the applied field.

The field dependence of the Mössbauer spectra of fast
frozen aqueous solutions of ferric salts is similar to
that of polycrystalline $Fe(NO_3)_3 \cdot 9H_2O$. Some results obtained for a 1 Mole solution of $FeCl_3$ at 78 K are shown in
Fig. 5. Fast freezing of aqueous solutions may lead to the
formation of a vitreous state in which the orientations
of the crystal field axes of neighbouring ferric ions are
more or less random. Therefore the observed field dependence of the spectra is probably due to an increase of the
spin-spin relaxation time as in $Fe(NO_3)_3 \cdot 9H_2O$. Heating of
fast frozen aqueous solutions above 200 K leads to an ir-
reversible transition to crystalline phases with shorter
relaxation time [23].

The field dependence of the spectrum of $Fe(acac)_3$ is
complicated. Small fields cause a decrease in the line
width, but larger fields broaden the line. A possible explanation is that two competitive effects are present,
namely a line narrowing as in Na_3FeF_6, $(NH_4)_3FeF_6$ and other
compounds, and an increase in the spin-spin relaxation time
as in $Fe(NO_3)_3 \cdot 9H_2O$. The latter effect is not unexpected,
as the space group of $Fe(acac)_3$ is Pbca and there are
eight molecules in the unit cell [24].

The spin-spin relaxation time is determined by H_{ex}
and H_{dd}. H_{ex} is often negligible when the interatomic distance between the ferric ions is large. Under these circumstances a variation in $<H_i>_{rms}$ should be reflected in
τ_{ss}. Some results are shown in Table 2.

The values of τ_{ss} for $(NH_4)_3FeF_6$ have an uncertainty
of about 20 % as the strength of the hyperfine interaction
is not known exactly. The values of τ_{ss} were determined

Table 2.   Spin-spin relaxation time estimated from
           Mössbauer spectra.

	$<H_i>_{rms}$	τ_{ss} (H=0 Oe)	τ_{ss} (H=6.5 kOe)
Ferric alum	450 Oe	$0.29 \cdot 10^{-9}$ sec	$2.0 \cdot 10^{-9}$ sec
$(NH_4)_3FeF_6$	1180 Oe	$0.12 \cdot 10^{-9}$ sec	$0.12 \cdot 10^{-9}$ sec

from the theories of Bradford and Marshall [4] and Wegener [2, 7] for H = 0 and H = 6.5 kOe, respectively. Apparently the dependence of τ_{ss} on $<H_i>_{rms}$ is more pronounced at H = 6.5 kOe than at H = 0 kOe. This may be due
to the fact that the electronic states are degenerate at
H = 0 kOe, whereas this is not the case at H = 6.5 kOe.
Therefore, different types of spin-spin relaxation processes may take place in the two cases.

8. SPIN-LATTICE RELAXATION

The spin-lattice relaxation time, τ_{sl}, is temperature
dependent due to the temperature dependence of the occupation of the phonon states. The temperature dependence of
spin-lattice relaxation can be studied by use of Möss-
bauer spectroscopy if $\tau_{sl} \gtrsim \tau_L$ and if $\tau_{ss} \gtrsim \tau_{sl}$. Furthermore
in order to obtain quantitative results, a theoretical
model describing the relaxation broadened spectra must be
available.

For temperatures well below the Debye temperature θ_D,
$\tau_{sl} \sim T^{-7}$, but for $T \gg \theta_D$, $\tau_{sl} \sim T^{-2}$ [25, 26]. Experimental
data only qualitatively agree with theoretical models for
spin-lattice relaxation [2, 26].

In $(NH_4)_3FeF_6$ the relaxation time is constant for temperatures up to 348 K, i.e. spin-lattice relaxation is slow compared to spin-spin relaxation.

We have studied the Mössbauer spectrum of $FeCl_3 \cdot 6H_2O$ above 78 K and have found that the line widths are independent of temperature, i.e. spin-lattice relaxation is too slow to be detectable. Even at 309 K, i.e. one degree below the melting point, no changes were found.

In several other compounds spin-lattice relaxation is detectable, but in most cases no quantitative information has been obtained.

9. DIMERS, TRIMERS AND SUPERPARAMAGNETISM

Several authors [27-30] have reported Mössbauer spectra of dimeric and trimeric ferric compounds. Dimeric ferric compounds with a strong antiferromagnetic coupling have a diamagnetic ground state. At low temperatures the Mössbauer spectra consist in symmetric quadrupole spectra with narrow lines. Contrary to monomeric compounds the application of a large magnetic field at 4.2 K does not result in a substantial magnetic splitting [28]. In $Fe(acac)_2Cl$ the finite size of the crystal field interaction causes a mixing of higher states into the ground state wave function. This leads to a temperature independent paramagnetism at low temperatures [27]. At high temperatures, where also the higher states are occupied, an asymmetric broadening of the quadrupole lines is often found in the spectra of dimeric compounds. This is due to relaxation between the excited paramagnetic states and the ground state.

Trimeric compounds have a paramagnetic ground state. Therefore, the application of large fields at 4.2 K results in substantial hyperfine fields [30]. The temperature dependence of the spectra is similar to that of dimeric compounds.

Clusters containing a large number of ferric ions with a strong exchange coupling often exhibit a superparamagnetic behaviour. However, a discussion of superparamagnetism is beyond the scope of this paper.

10. CONCLUSION

Detailed information on the crystal field interaction in paramagnetic compounds can be obtained from Mössbauer spectra when the relaxation time is long compared to the nuclear Larmor period. However, this is rarely the case in pure compounds and the investigations are in practice limited to the study of ferric ions as impurities in diamagnetic matrices.

In this paper we have shown that information about the crystal field interaction and the relaxation processes can be obtained from the relaxation broadened spectra of some pure compounds. Many of the results are only qualitative, and more theoretical work is required in order to obtain quantitative information. However, a large number of unknown parameters are involved. Therefore, an unambiguous interpretation of the spectra may not always be practicable.

ACKNOWLEDGEMENT

The author is grateful to Professor G. Trumpy for his continuous interest in this work and to Dr. N. Thrane who has made contributions to this work and has constructed most of the experimental set-up.

1. S. Mørup and N. Thrane, in "Hyperfine Interactions in Excited Nuclei", edited by G. Goldring and R. Kalish (Gordon and Breach, New York, 1971), p. 827.

2. S. Mørup and N. Thrane, Phys. Rev. B $\underline{4}$, 2087 (1971).

3. S. Mørup and N. Thrane, Phys. Rev. B $\underline{8}$, 1020 (1973).

4. E. Bradford and W. Marshall, Proc. Phys. Soc. (London) $\underline{87}$, 731 (1966).

5. A. M. Afanas'ev and V. D. Gorobchenko, Report IAE-2215, Moscow, 1972.

6. F. van der Woude and A. J. Dekker, Phys. Stat. Sol. $\underline{9}$, 775 (1965).

7. H. Wegener, Z. Physik $\underline{186}$, 498 (1965).

8. M. Blume, Phys. Rev. Letters $\underline{14}$, 96 (1965).

9. M. Blume and J. A. Tjon, Phys. Rev. $\underline{165}$, 446 (1968).

10. H. H. Wickman, M. P. Klein, and D. A. Shirley, Phys. Rev. $\underline{152}$, 345 (1966).

11. M. D. Lind, J. Chem. Phys. $\underline{47}$, 990 (1967).

12. M. Kaplan and T. X. Carroll, in Proceedings of the Conference on the Application of the Mössbauer Effect,

Tihany, 1969, edited by I. Dézsi (Académiai Kiadó, Budapest, 1971), p. 169.

13. N. Thrane, in Proceedings of the Conference on the Application of the Mössbauer Effect, Tihany, 1969, edited by I. Dézsi (Académiai Kiadó, Budapest, 1971), p. 175.

14. N. Thrane and G. Trumpy, Phys. Rev. B $\underline{1}$, 153 (1970).

15. M. Blume, Phys. Rev. Letters, $\underline{18}$, 305 (1967).

16. H. H. Wickman and C. F. Wagner, J. Chem. Phys. $\underline{51}$, 435 (1969).

17. M. D. Lind, M. J. Hamor, T. A. Hamor, and J. L. Hoard, Inorg. Chem. $\underline{3}$, 34 (1964).

18. H. Bode and E. Voss, Z. Anorg. Allgem. Chem. $\underline{290}$, 1 (1957).

19. A. M. Afanas'ev and Yu. M. Kagan, Zh. Eksperimen. i Teor. Fiz. Pis'ma v Redaktsiyu $\underline{8}$, 620 (1968). [Sov. Phys. JETP Letters $\underline{8}$, 382 (1968)].

20. V. D. Gorobchenko, I. I. Lukashevich, V. V. Sklyarevskii, K. F. Tsitskishvili, and N. I. Filippov, Zh. Eksperim. i Teor. Fiz. Pis'ma v Redaktsiyu $\underline{8}$, 625 (1968) [Sov. Phys. JETP Letters $\underline{8}$, 386 (1968)].

21. S. Mørup and N. Thrane, Chem. Phys. Letters $\underline{21}$, 363 (1973).

22. S. Mørup, J. Phys. Chem. Solids (in press).

23. I. G. Gusakovskaya, T. I. Larkina, and V. I. Goldanskii, Fizika Tverdogo Tela, $\underline{14}$, 2631 (1972) [Sov. Phys. Solid State $\underline{14}$, 2274 (1973)].

24. R. B. Roof, Acta Cryst. $\underline{9}$, 781 (1956).

25. V. V. Svetozarov, Fizika Tverdogo Tela, 12, 1054 (1970) [Sov. Phys. Solid State 12, 826 (1970)]. ·

26. I. P. Suzdalev, A. M. Afanas'ev, A. S. Plachinda, V. I. Gol'danskii, and E. F. Makarov, Zh. Eksp. Teor. Fiz., 55, 1752 (1968) [Sov. Phys. JETP 28, 923 (1969)].

27. B. W. Fitzsimmons and C. E. Johnson, Chem. Phys. Letters 6, 267 (1970).

28. A. N. Buckley, I. R. Herbert, B. D. Rumbold, and G. V. H. Wilson, J. Phys. Chem. Solids, 31, 1423 (1970).

29. A. N. Buckley, B. D. Rumbold, G. V. H. Wilson, and K. S. Murray, J. Chem. Soc. (A), 1970, 2298.

30. B. D. Rumbold and G. V. H. Wilson, J. Phys. Chem. Solids 34, 1887 (1973).

PHASE TRANSITIONS IN ANTIFERROMAGNETS IN EXTERNAL MAGNETIC FIELDS: MÖSSBAUER SPECTROSCOPY

R. B. Frankel
Francis Bitter National Magnet Laboratory*
Massachusetts Institute of Technology
Cambridge, Massachusetts 02139

I. INTRODUCTION

Known antiferromagnetic materials far out number known ferromagnetic and ferrimagnetic materials. Moreover, the antiferromagnets display a wide variety of phase transitions in addition to the paramagnetic to antiferromagnetic transition at the Néel point, that make them a rich source of information about the relationship between exchange interactions, structure, and phase transition phenomena. Mössbauer spectroscopy has been used extensively over the last decade to study antiferromagnets, capitalizing on the fact that the magnetic hyperfine field is proportional to the sublattice magnetization. In general, these studies have concentrated on determining the magnetic structure below T_N and the sublattice magnetization as a function of T below T_N, in some cases in the critical region. Less well studied are those transitions in antiferromagnets which are induced by external magnetic fields, although in these cases too, Mössbauer spectroscopy can provide important and often unique information.

In this paper we will review some applications of Mössbauer spectroscopy to the study of phase transitions in antiferromagnets induced by external magnetic fields. In Sec. II below we will briefly recapitulate the molecular field approximation (MFA) and review the phase transitions in antiferromagnets; in Sec. III we will make some general

151

remarks about Mössbauer spectroscopy of these transitions; in Sec. IV we will briefly review some applications to specific materials.

II. PHASE TRANSITIONS IN SIMPLE UNIAXIAL ANTIFERROMAGNETS

A. Phases and Phase Boundaries [1-5]

The simplest kind of antiferromagnet consists of two interpenetrating sublattices, 1 and 2, where the spin moment of a magnetic ion at any given site is antiferromagnetically coupled to the spin moments of its neighboring ions. Thus at temperatures below T_N and external magnetic field $H_o = 0$, the sublattice moments, σ_1 and σ_2 are antiparallel to each other. The orientation of σ_1 and σ_2 relative to the crystal structure of the material is determined by the so-called anisotropy energy. If the anisotropy energy has uniaxial symmetry and is of such magnitude and sign such that the axis of symmetry (say the a-axis) is the preferred direction of σ_1 and σ_2, then the antiferromagnet is called uniaxial and of the easy-axis type.

Below T_N, a critical value of an external magnetic field $H_o = H_{SF}$ applied parallel to the a-axis will induce a first-order-phase transition to a phase in which σ_1 and σ_2 are roughly perpendicular to each other and perpendicular to the a-axis (in the basal plane). This phase is known as the spin-flop phase (SF). If H_o is increased beyond H_{SF} then σ_1 and σ_2 tip toward each other and the a-axis until a second critical value of the external field $H_o = H_P$ induces a second-order-phase transition to the paramagnetic phase (P) in which σ_1 and σ_2 are parallel to each other and parallel to the a-axis.

Both H_{SF} and H_P are determined by the exchange energy and the anisotropy energy (see below), and as both are temperature dependent, the critical magnetic fields H_{SF} and H_P are also temperature dependent, and a representative phase diagram of a uniaxial, easy-axis antiferromagnet is shown in Fig. 1a. There is a triple point (H_3, T_3) where all three phases coexist; for $T > T_3$, only the AF and P phases exist.

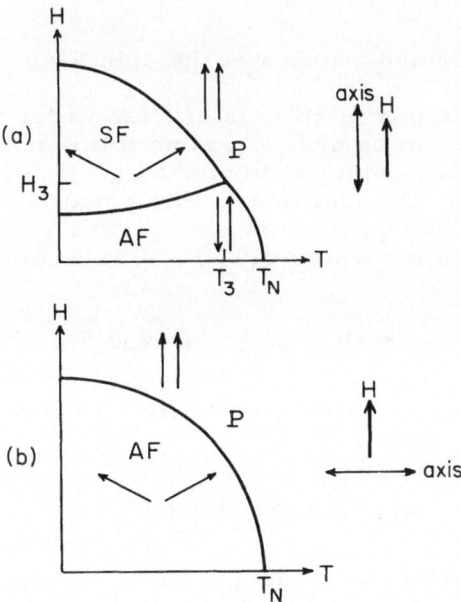

Fig. 1. Schematic phase diagram for uniaxial, easy-axis antiferromagnet: a) H_o parallel to the easy axis; b) H_o perpendicular to the easy axis (after Ref. 2).

A magnetic field applied perpendicular to the easy axis for $T < T_N$ will induce a phase transition from the AF phase to the P phase, as shown in Fig. 1b. In this case, the field causes the spins to tip toward each other and away from the a-axis until the critical field H_{AFP} causes the transition to the P phase, where they are parallel to each other and perpendicular to the a-axis. There is no SF phase for the orientation of H_o relative to the a-axis.

B. Thermodynamics of the Spin Flop

The spin-flop transition is of first-order thermodynamically and can be understood as follows.[5] As the analog of the Gibbs potential for nonmagnetic systems, we define $\Phi(T, P, H)$, the thermodynamic potential for the antiferromagnet. The equilibrium configuration of spins is that configuration for which $\Phi(T, P, H)$ is a minimum. At constant T and P,

$$\Phi(H) = \Phi(0) - \frac{1}{2} \chi H_o^2 \tag{1}$$

where χ is the magnetic susceptibility. At $H = 0$, $\Phi_{AF}(0) < \Phi_{SF}(0)$ and

$$\Phi_{SF}(0) = \Phi_{AF}(0) + K, \tag{2}$$

where K is the anisotropy energy. The critical field H_{SF} is by definition that field for which the AF and SF phases coexist, or

$$\Phi_{AF}(H_{SF}) = \Phi_{SF}(H_{SF}). \tag{3}$$

Using Eqs. (1) and (2), Eq. (3) leads to

$$H_{SF} = \left[\frac{2K}{\chi_{SF} - \chi_{AF}} \right]^{1/2} \tag{4}$$

Since $\chi_{AF} = \chi_{\parallel}$, the low field susceptibility with H_o oriented along the a-axis, and $\chi_{SF} \approx \chi_{\perp}$, the low field susceptibility with H_o oriented perpendicular to the a-axis, Eq. (4) can be rewritten as

$$H_{SF} = \left[\frac{2K}{\chi_{\perp} - \chi_{\shortparallel}} \right]^{1/2}. \tag{5}$$

$\chi_{\shortparallel}$ approaches zero at T goes to zero while $\chi_{\perp}$ is roughly constant below T_N for an ideal, uniaxial antiferromagnet (see Fig. 2). Above T_N $\chi_{\shortparallel} \approx \chi_{\perp}$. Moreover, $\chi_{\perp} - \chi_{\shortparallel}$ usually decreases faster than K so that H_{SF} increases with increasing T.

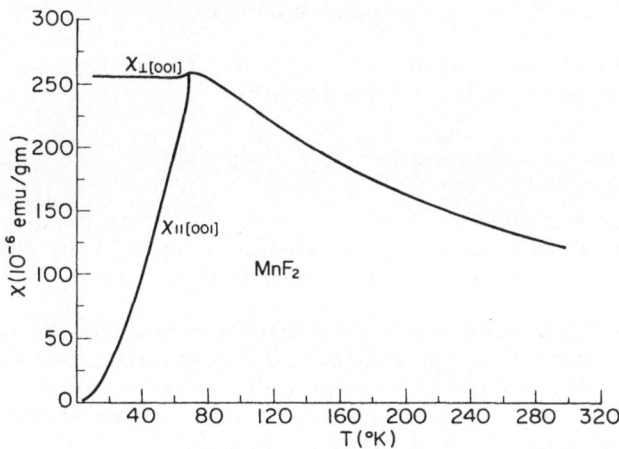

Fig. 2. $\chi_{\shortparallel}$ and $\chi_{\perp}$ plotted as a function of temperature for the ideal uniaxial antiferromagnet MnF_2 (after Ref. 1).

C. Molecular Field Approximation

If we assume that the ions on one sublattice interact only with the neighboring ions on the other sublattice, then the molecular field approximation (MFA) consists in replacing the exchange interaction by an effective exchange field which is proportional to the magnetization of the other sublattice. If the spin of the transition metal ion is S, then the magnetization $\sigma = <S>/S$ is given by a Brillouin function

$$\sigma_1 = B_s \left[g\mu_B S \frac{H_1}{kT} \right], \tag{6}$$

where H_1 is the total field given by

$$H_1 = H_o + H_{ex} = H_o - \lambda \sigma_2. \tag{7}$$

The expressions for σ_2 are similar. The constant λ includes the number of neighbors and the exchange integral.[2,3,4]

In the MFA, the AF to P phase boundary is given by[2]

$$H_{AFP} = 2\lambda\sigma(T) \tag{8}$$

where $\sigma(T)$ is calculated with $H_o = 0$. Close to T_N, this leads to an expression of the form[2,4]

$$T_N - T = D\, H_{AFP}^2, \tag{9}$$

where D is a constant, i.e., that the depression of the Néel point is quadratic in magnetic field. For $H_o \perp a$, the phase boundary has the same form, with $D(H_o \perp a) \approx 1/3\, (D(H_o \parallel a).$[2]

The MFA predicts a similar form for the SF to P phase boundary, except that small corrections for the anisotropy energy must be included, which change the extrapolated zero field transition temperature from T_N. More sophisticated calculations predict

$$H_P(T) = H_P(0) \left[1 - aT^{3/2} \right], \tag{10}$$

for $T \to 0$, where a is a constant.

The MFA can also be used to calculate the individual sublattice magnetizations for $H_o \parallel a$ and $T_3 < T \leq T_N$ and $T > T_N$.[2,4] Although $\sigma_1 = -\sigma_2$ for all T when $H_o = 0$, in finite H_o this is no longer the case. Moreover $\sigma_1 = \sigma_2 \neq 0$ for $T > T_N$ if $H_o \neq 0$, but $\sigma_1 \neq -\sigma_2$ for $T < T_N$. The expected sublattice magnetizations as a function of H_o are indicated schematically in Fig. 3. The sublattice magnetizations as function of T at constant H_o have a similar form.

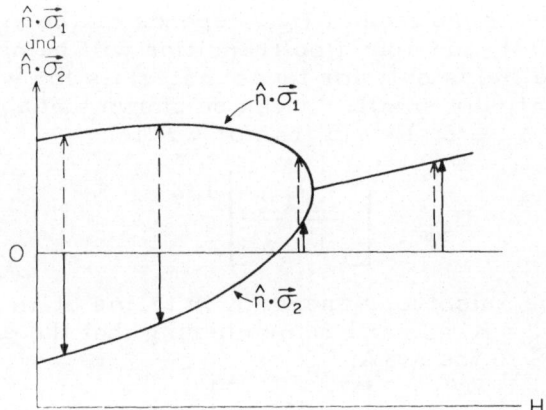

Fig. 3. Magnetic field dependence of $\hat{n} \cdot \vec{\sigma}_1$ and $\hat{n} \cdot \vec{\sigma}_2$ at constant T, $T_3 < T < T_N$. $\hat{n}$ is a unit vector pointing parallel to $\vec{H}_o$ and along the easy axis. The solid and dashed arrows represent $\vec{\sigma}_1$ and $\vec{\sigma}_2$, respectively (after Ref. 2).

It is well known that the MFA breaks down in the critical region, i.e., close to T_N. Here it is more appropriate to use this expression

$$\sigma_i = A(1 - T/T_N)^\beta, \qquad (11)$$

where A is a constant and β is a constant, usually 0.33 for a three-dimensional antiferromagnet. This expression is also suitable for measurements in external magnetic fields, as is shown below in Sec. IV-A. In external fields however, one uses

$$\vec{L} = \frac{\vec{\sigma}_1 - \vec{\sigma}_2}{2},$$

to compare with the second member of Eq. (11).

Since the magnitude of H_{SF} depends on the anisotropy energy (Eq. (5)), the spin-flop transition will be observable with available fields only for those materials for which this energy is relatively small. It can be shown that $\chi_\perp \approx 1/\lambda$ and taking $\chi_\parallel / \chi_\perp \approx \delta$, Eq. (5) can be written[1]

$$H_{SF} = \left[\frac{2\lambda K}{1 - \delta} \right]^{1/2} . \tag{12}$$

If we write the anisotropy energy K in terms of an aniso-tropy field H_A = K/gS, and remembering that the exchange field H_E = λgS in the MFA,

$$H_{SF} = \left[\frac{2H_E H A}{1 - \delta} \right]^{1/2} . \tag{13}$$

Since $\delta \approx 0$ at T = 0, it can be seen that for exchange fields of 10^6 Oe, the anisotropy field must be less than 10^4 Oe in order that the spin flop be observable in the laboratory ($H_0 \approx 2 \times 10^5$ Oe).

D. More Complex Systems

Ideal situations are encounted infrequently in any area of experience, and so too with antiferromagnets. It would not be possible to recapitualte all the possible com-plications, however, one general type of transition should be mentioned. In the simple MFA above, we assumed that the entire exchange field acting on a given site was due to the neighboring ions on the other sublattice. An obvious extension of the model would be to include interactions of the ion with other ions on the same sublattice. In some cases the structure is such that the intra-sublattice inter-action is stronger than the inter-sublattice interaction, i.e., the anisotropy field is effectively greater than the anti-ferromagnetic exchange field. If a field is applied along the easy axis, it may be possible to observe a first-order AF to P transition, called a metamagnetic transition. In certain cases, with more complex structures, intermediate ferrimagnetic phases can be observed with increasing H_0 along a, all first-order thermodynamically.

III. HYPERFINE INTERACTIONS AND MÖSSBAUER SPECTROSCOPY

Associated with the phase boundaries outlined in the previous sections, there will be changes in the hyperfine spectrum which can be observed by Mössbauer spectroscopy. The general changes which one might observe are: a) the magnitude of the hyperfine interaction changes in going across the phase boundary with consequent shifts in the spectral lines; b) the polarization direction changes, with consequent changes in the relative intensities of the lines; c) the angle between the principle component of the electric field gradient and the magnetic hyperfine field changes with consequent shifts in the spectral lines.

Let us consider the expected changes with reference to a uniaxial antiferromagnet with H_0 along the easy (a) axis (see Fig. la) and parallel to the γ-ray propagation direction. Let us further assume that V_{zz} is along a and $\eta = 0$. Then along the $H_0 = 0$ line, above T_N one obtains a quadrupole spectrum only, while below T_N one obtains a single spectrum with the angle β between V_{zz} and $\vec{H}_{hf} = 0$. Of course, $\vec{H}_{hf}$ is temperature dependent, going to zero at T_N. For a single crystal absorber, $\Delta m = 0$ lines have intensity zero.

A. AF Region

Since the total field at the nucleus is the sum of the hyperfine field and the applied field,

$$\vec{H}_n = \vec{H}_{hf} + \vec{H}_o, \qquad (14)$$

application of H_0 along a results in two spectra, one in which $H_n = H_{hf} + H_0$, the other in which $H_n = H_{hf} - H_0$. Unless the sign of the hyperfine interaction is known (from measurements in the P phase, for example) it is not possible to say which spectrum corresponds to the spin up sublattice, and which corresponds to spin down. For both spectra, β remains zero and the intensity of the $\Delta m = 0$ lines equals 0. In crossing the AF-P phase boundary, H_{hf} will be proportional to σ and can be compared with the MFA calculations outlined above. Since both sublattices will, in general, have different magnetizations σ, so will the hyperfine fields for the ions on the two sublattices be different. In the critical region it is appropriate to take the vector difference of the two hyperfine fields $\vec{H}_{hf}^{(1)} - \vec{H}_{hf}^{(2)}$ (not $\vec{H}_n^{(1)} - \vec{H}_n^{(2)}$) proportional to $\vec{L}$ to compare with Eq. (11).

B. SF Region

At $H_o = H_{SF}$, the two spectra collapse into a single spectrum with the $\Delta m = 0$ lines now the most intense. The angle β is now $\pi/2$, and $H_n \approx \sqrt{H_o^2 + H_{hf}^2}$. In addition, the magnitude (and possibly even the sign) of H_{hf} in the SF region, say H_{hf}^{SF}, may be different than in the AF region, say H_{hf}^{AF}. In Fe^{3+}, the change in H_{hf} is not expected to be large because Fe^{3+} is an S-state ion, but the change can be considerable in Fe^{2+} where the single ion anistropy can be large. If the spin-flop boundary is at large values of H_o, then the spins in the SF phase will be canted out of the basal plane toward a. This means that β will be less than $\pi/2$ and the appropriate vector sum of $\vec{H}_{hf}$ and $\vec{H}_o$ must be taken to give $\vec{H}_n$.

C. P Region

In the P region, after crossing either the SF-P phase boundary or the AF-P phase boundary, the spectrum is again a single spectrum with $H_n = H_{hf} \pm H_o$, depending on the sign of H_{hf}. Again, as in the AF region, $\beta = 0$ and the $\Delta m = 0$ lines have zero intensity. The magnitude of H_{hf} is H_o and T dependent, and the dependence can usually be approximated by a Brillouin function,

$$H_{hf} = H_{hf}^0 \, B_s \left(\frac{g \mu_B \, H_i}{kT} \right), \tag{15}$$

where H_{hf}^0 is the saturation value of the hyperfine interaction, i.e., the value of H_{hf} for $B_s = 1$, and H_i is the magnetic field acting on the ionic moment, including H_o and the exchange fields due to the other spins (Eq. (7) above).

D. H_o Applied Perpendicular to the a-axis

In this case we refer to the phase diagram in Fig. 1b. In the AF region, one now obtains a single spectrum for all

H_o with $H_n \approx \sqrt{H_o^2 + H_{hf}^2}$ and $\beta = 0$. If the γ-ray propagation direction is parallel to a, then the $\Delta m = 0$ lines will be absent. As H_o increases and the spins tip away from a, β will decrease and the $\Delta m = 0$ lines will increase in intensity. At the AF-P phase boundary H_{hf} and H_o are parallel, $\beta = \pi/2$ and the $\Delta m = 0$ lines have maximum intensity. The value of H_{hf} in the P phase may be different than that observed in the P phase with $H_o \parallel a$, reflecting anisotropy in the single ion properties.

IV. EXPERIMENTAL RESULTS

A. MnF_2

MnF_2 crystalizes in a rutile structure with a tetragonal lattice and two magnetic ions per unit cell. Below the Néel temperature $T_N = 67.4$ K, the magnetic properties of MnF_2 are understood in terms of an ideal, two sublattice, easy-axis antiferromagnet with the spins aligned along the c-axis. The phase diagram of MnF_2 has been studied by magnetic moment and ultrasonic techniques[2] and is shown in Fig. 4. H_{SF} is 93 kOe at 4.2 K; this low field is due to the fact that Mn^{2+} is an S-state ion and has low anisotropy.

Fe^{2+} may be isomorphously incorporated into the MnF_2 lattice, and zero field Mössbauer spectroscopy has been reported by Wertheim et al.[6] The addition of iron results in an increase of the Néel temperature and an increase in the value of H_{SF}, due to the fact that Fe^{2+} has a large single ion anistropy compared with Mn^{2+}. For 1% Fe in MnF_2, H_{SF} is 105 kOe.[7]

Spin Flop. The SF phase was observed by Abeledo et al.[8] using Mössbauer spectroscopy. A large single crystal of ~1% Fe^{2+} doped MnF_2 was grown from the melt by Optovac, Inc. The crystal was oriented and a 6 mil slice was taken perpendicular to the c-axis. H_o was applied parallel to the c-axis.

Spectra at 4.2 K and $H_o = 0$, 75 and 111 kOe are shown in Figs. 5a, 5b and 6 respectively. For $H_o = 0$ an apparent three line spectrum is observed, due to a fortuitous super-position of the inner $\Delta m = \pm 1$ lines. In the AF phase, $H_{hf}^{AF} = 227$ kOe, $\Delta E = 2.8$ mm/sec and $\beta = \pi/2$.[6] For $H_o < H_{SF}$, the spectrum (Fig. 5b) consists of the super-

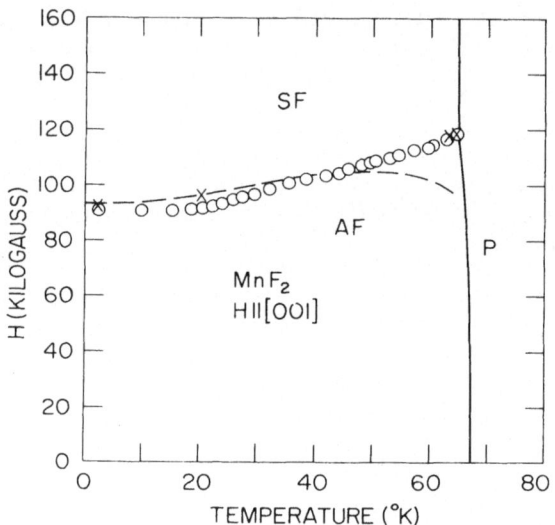

Fig. 4. AF-SF phase boundary in MnF_2, with H_O parallel
to $[00\bar{1}]$ (after Ref. 2).

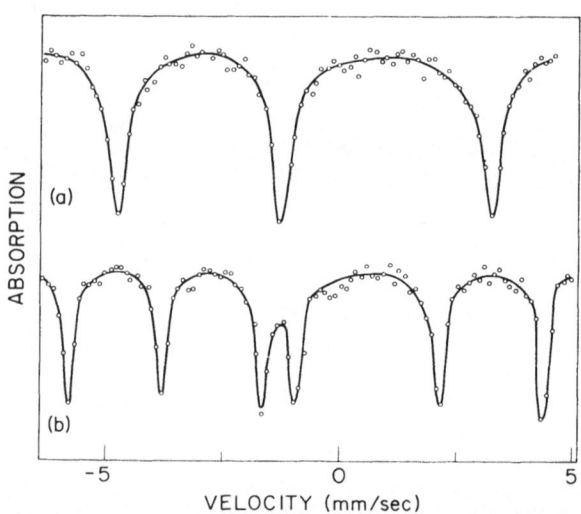

Fig. 5. Mössbauer spectra of single crystal $MnF_2{:}Fe^{2+}$:
a) 4.2 K, $H_O = 0$; b) 4.2 K, $H_O = 75$ kOe; $\gamma \parallel H_O \parallel [00\bar{1}]$.

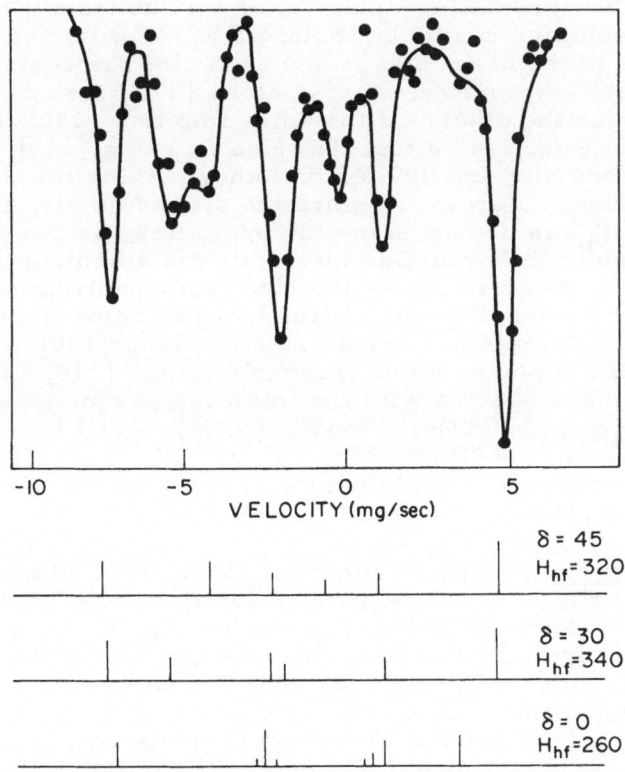

Fig. 6. Mössbauer spectrum of single crystal MnF$_2$ at 4.2 K, H$_0$ = 111 kOe, in the SF phase. The bar diagrams below the figure represent the expected spectra for various orientations of the moment in the basal plane (after Ref. 8).

position of the two sublattice spectra as discussed in Sec.III.A.
For $H_o > H_{SF}$, the spectrum (Fig. 6) changes dramatically
due to the appearance of the $\Delta m = 0$ lines, and the field at
the nucleus $\vec{H}_n = \vec{H}_o + \vec{H}_{hf}^{SF}$, where H_{hf}^{SF} is perpendicular to c.
The spectrum is considerably complicated by the fact that
there are two crystallographic sites for the transition metal
ions with common c-axis but oriented at 90° with respect to
each other in the basal plane, and each site has equivalent
orthorhombic symmetry. V_{zz} is along $[110]$ for one site
and $[1\bar{1}0]$ for the other. If the spins flop to a $[100]$ direction,
β is the same for both sites. If spins flop to a $[110]$ direction,
$\beta = 0^o$ for one site and 90° for the other, giving two superposed spectra. Moreover, since the site symmetry is orthorhombic, H_{hf}^{SF} is not the same for both sites, as the orbital
contribution to the hyperfine interaction is anisotropic in the
basal plane. Comparison of the observed spectrum with
computer calculated spectra leads to a two domain model,
in which one domain has spins oriented along $[100]$, and
the other has spins oriented (probably) along $[110]$. Hence
there are three spectra with the following parameters: i)
H_{hf}^{SF} ($[100]$) = 320 kOe, $\beta = 45^o$, ii) H_{hf}^{SF} ($[110]$) = $- 340$ kOe,
$\beta = 90^o$, iii) H_{hf}^{SF} ($[110]$) = 260 kOe, $\beta = 0^o$. These results
have been analyzed to yield values of the g-factor of Fe^{2+}
in the basal plane. For details see Ref. 8.

AF to P.[9] The phase diagram close to T_N is shown in
Fig. 7. In Fig. 8 we show spectra for $H_o = 80$ kOe along the
c-axis both above and below T_N. Below T_N the spectrum
consists of eight lines, due to the superposition of two four-
line spectra; above T_N the spectrum has collapsed to a single
four-line spectrum. Since the field at the nucleus H_n is less
than H_o, H_{hf}^P is negative. The field dependence of the phase
boundary was determined by sweeping the temperature at
constant field and the field at constant temperature, and is
shown in Fig. 9, where H_o^2 at T_N is plotted as a function of
T. The straight line corresponds to Eq. (9); also shown is
corresponding phase boundary for pure MnF_2.[2]

In Fig. 10, the hyperfine field for each sublattice is
shown plotted as a function of T for $H_o = 80$ kOe. The dashed
line is the best fit with the MFA above and below T_N, and the
solid line is the best fit with the MFA to the Néel point itself.
In this case the MFA is more complicated than outlined in
Sec. II.C, because it is necessary to account for the Mn-Mn
interaction, the Fe-Fe interaction, as well as the Mn-Fe interaction. Moreover, it is necessary to take the Fe^{2+} fine structur

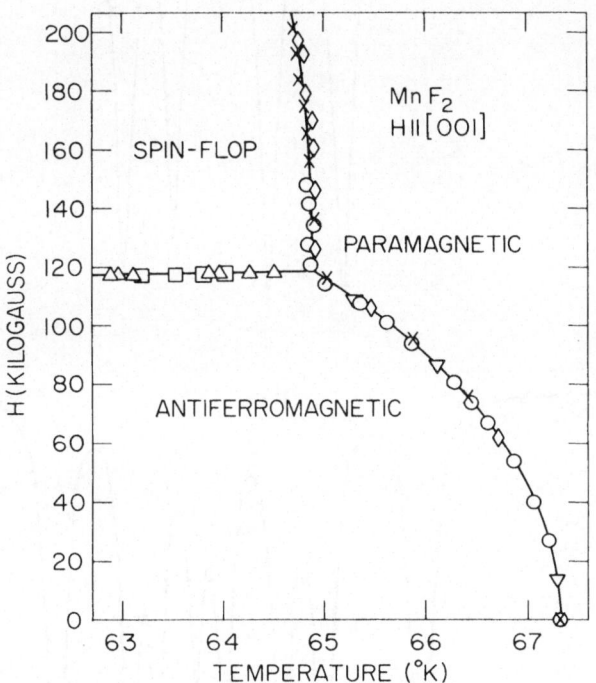

Fig. 7. Phase diagram of MnF_2, H ∥ [001], close to T_N (after Ref. 2).

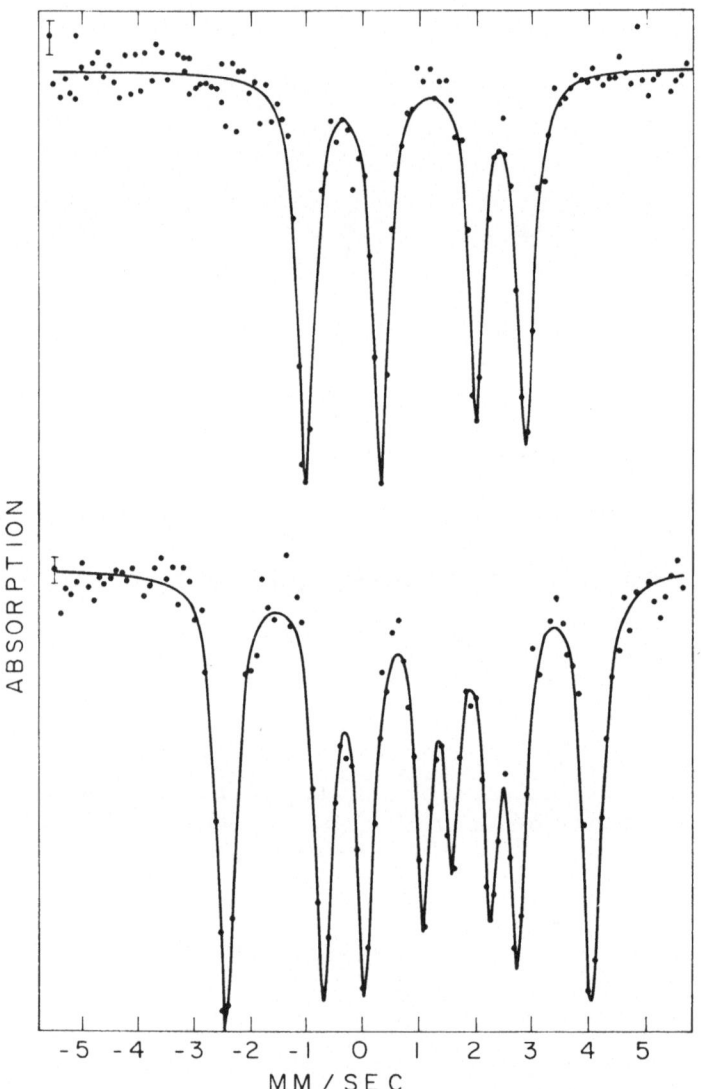

Fig. 8. Spectra of single crystal $MnF_2:Fe^{2+}$ at 80 kOe parallel to [001] above (top spectrum) and below (bottom spectrum) T_N (after Ref. 9).

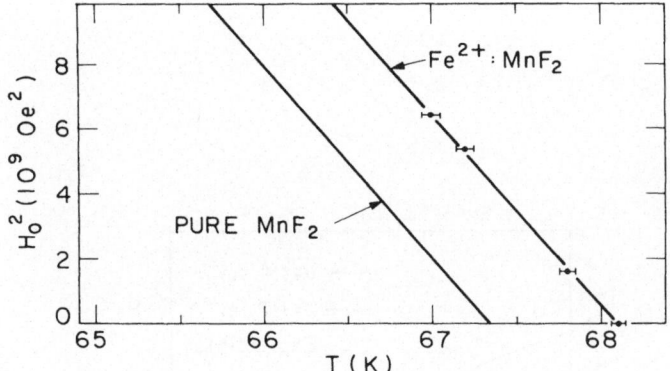

Fig. 9. H_o^2 at the AF-P phase boundary plotted as a function of T for MnF_2 and $MnF_2:Fe^{2+}$ ($H_o \parallel [001]$) (after Ref. 9).

splitting into account. In the calculations shown in Fig. 10, J(Mn-Mn) and J(Fe-Fe) were taken from the Néel temperatures of pure MnF_2 and Fe F_2 respectively, using the relation

$$J = \frac{3 k \ T_N}{2zS (S + 1)}$$

where z is the number of nearest neighbors. This reduces the number of parameters in the MFA to two, i.e., J(Mn-Fe) and the Fe^{2+} fine structure splitting. The latter can be obtained from the shift of H_{SF} with concentration in $Mn_x Fe_{1-x} F_2$ crystals.[7] The value of the Mn-Fe exchange obtained,[9]

$$J(Mn-Fe) = - 1.7 \ cm^{-1},$$

is in fair agreement with determinations by other methods.

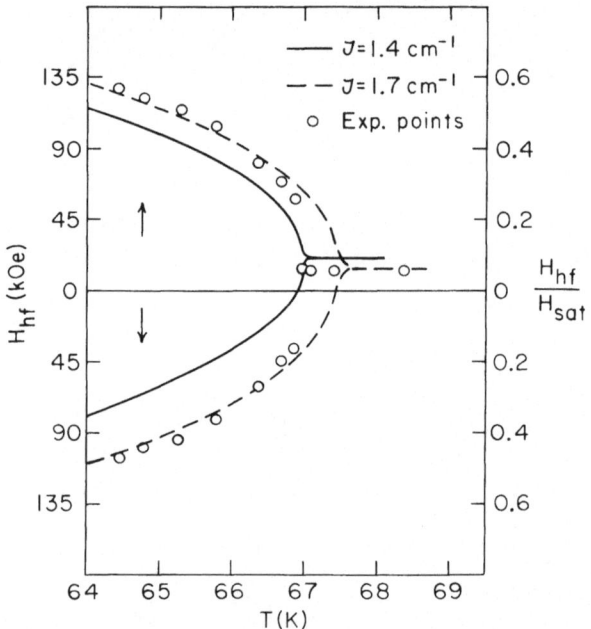

Fig. 10. $|H_{hf}|$ and $\sigma = H_{hf}/H_{sat}$ plotted as a function of T for $H_O = 80$ kOe, $H_O \parallel [001]$. The solid line is the best fit with the MFA, above and below T_N. The dashed line gives the best fit to the Néel temperature.

Critical Region. As explained in Sec. II, the best fit with the MFA breaks down near T_N. In Fig. 11 we show log L plotted as a function of log $(1 - T/T_N)$. Data for $H_o = 0$ and 80 kOe are shown plotted together with data obtained at H = 0 by Wertheim et al.[6] As can be seen, the same critical exponent $\beta = 0.334$ fits the data over the wide range of magnetic field, from 0 to 80 kOe.

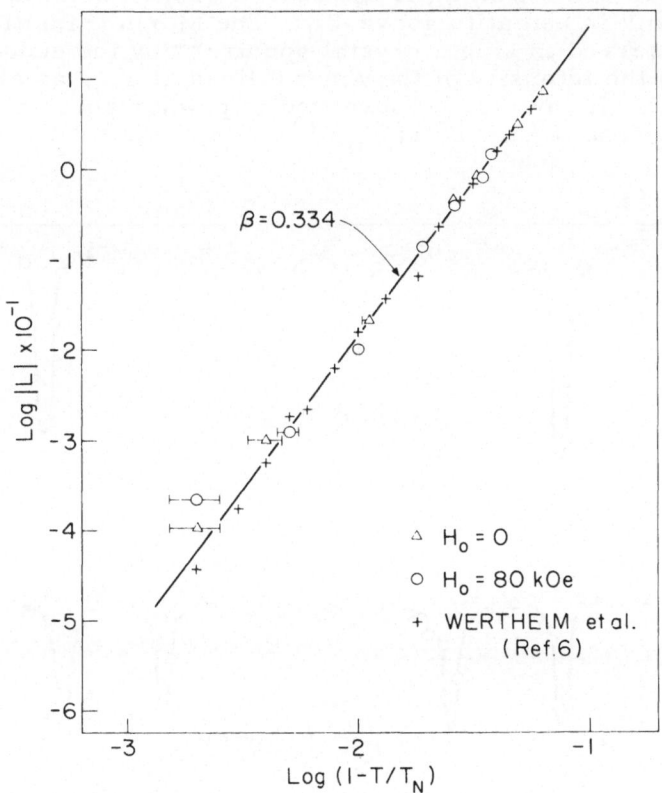

Fig. 11. Log L plotted as a function of Log $\left| T - T_N \right|$, for $H_o = 0$ and 80 kOe, with $H_o \parallel [001]$.

B. α-Fe$_2$O$_3$

Hematite (α-Fe$_2$O$_3$) is essentially antiferromagnetic below the Néel temperature $T_N \approx 960$ K. [10, 11] The magnetic structure is more complicated than the simple uniaxial anti-ferromagnet because the anisotropy, while small, is tempera-ture dependent and in fact changes sign, leading to a spontan-eous spin flip in zero magnetic field known as the Morin transition, with $T_m \approx 260$ K. For $T < T_m$, the spins are aligned antiparallel along the trigonal axis. For $T_m < T < T_N$, the spins lie in the basal plane, and are slightly canted toward each other. This canting is the source of the "weak ferro-magnetism" in hematite above T_N. The Morin transition is easily observed in single crystal spectra[12] by the sudden change in the intensity of the $\Delta m = 0$ lines at T_m, as shown in Fig. 12. It can also be observed in powder spectra because β changes from $0°$ to $90°$ at T_m. [13]

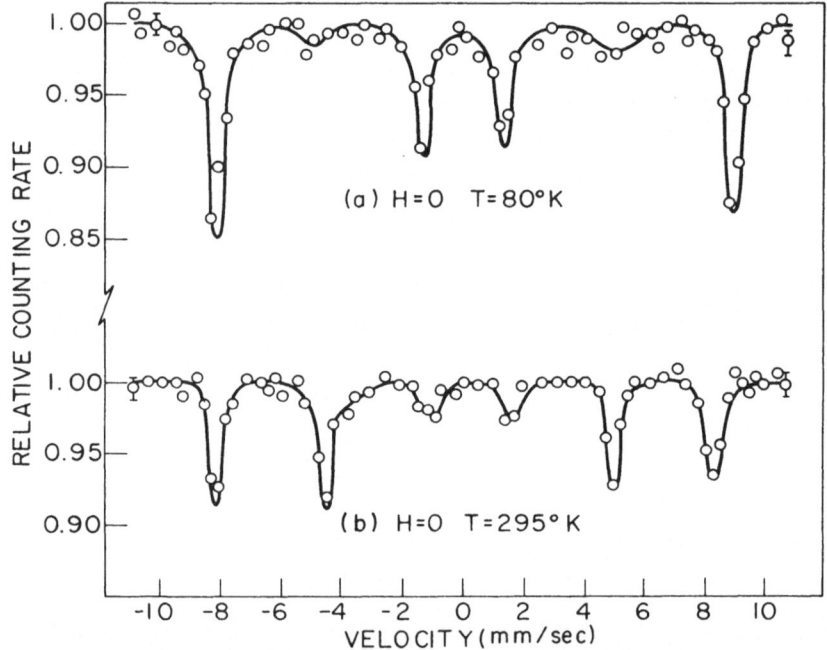

Fig. 12. Mössbauer spectra of single crystal α-Fe$_2$O$_3$ with $\gamma \parallel$ c-axis, above and below T_m (after Ref. 12).

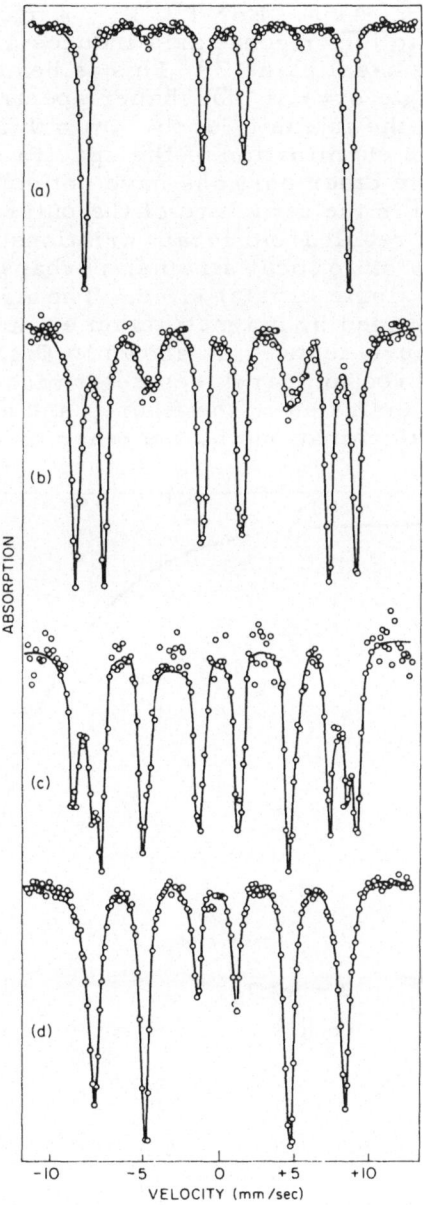

Fig. 13. Spectra of single crystal α-Fe_2O_3 at 4.2 K with $H_o \parallel c$: a) $H_o = 0$; b) $H_o = 65$ kOe; c) $H_o = 66.5$ kOe; d) $H_o = 70$ kOe.

H_o ‖ Trigonal Axis.   For $T < T_m$, a magnetic field
applied parallel to the trigonal axis induces a first-order
spin flop into the basal plane.[12]   This is beautifully illustrated by the single crystal Mössbauer spectra in Fig. 13.
At the spin flop, the intensity of the $\Delta m = 0$ lines changes
as does β.   Close examination of the spectra show that portions
have flopped while other portions have not yet flopped; this is
most easily seen in the structure of the outer lines.   These
"domains" could result from local variations in the anisotropy
due to impurities or to local strains, perhaps introduced in
producing a thin single crystal slice.   The AF-SF phase
boundary, determined by magnetic moment and ultrasonic
attenuation measurements[14] is shown in Fig. 14.   The transition fields at three temperatures determined by the Mössbaue
effect[12, 15] are indicated in the figure, and are in good agreement with the determinations by the other methods.

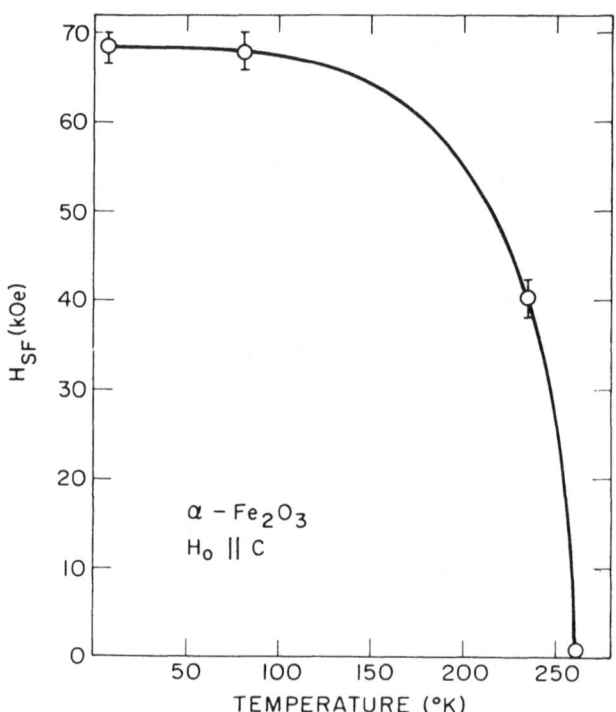

Fig. 14.   Phase boundary of α-Fe_2O_3 (after Ref. 14).   The
data points are spin-flop fields determined by the Mössbauer
spectra.

$H_o \perp$ Trigonal Axis. Because of the complicated anisotropy, H_o applied perpendicular to the trigonal axis will also
induce a first-order spin flop.[14] This was studied close to
T_m by Simkin and Bernheim[16] using the Mössbauer effect,
and at lower temperature by Blum and Frankel.[15] The
complete phase boundary was determined by ultrasonic and
magnetic moment measurements.[14] Blum and Frankel[15]
found from measurements of the intensity of the $\Delta m = 0$ lines
as a function of H_o, that except close to T_m, the spins rotate
away from the trigonal axis toward the basal plane before
flopping. The rotation is large enough that the first-order
transition could not be observed with the Mössbauer effect,
except just below T_N.[15]

C. Metamagnetic Transitions and Complex Structures

FeCl$_2$· 2H$_2$O. Monoclinic $FeCl_2$ · $2H_2O$ orders antiferromagnetically at $T_N \approx 23$ K and the magnetic structure
consists of two sublattices of -FeCl$_2$-chains lying along the
c-axis.[17] The coupling along the chains is ferromagnetic
with weak antiferromagnetic coupling between chains. Application of an external magnetic field along the easy axis a induces phase transitions at $H_1 = 39$ kOe and $H_2 = 46$ kOe.[17, 18]

Mössbauer studies of $FeCl_2$·$2H_2O$ include a powder study
by Chandra and Hoy[19] and a single crystal study by Johnson.[20]
They found a magnetic hyperfine field of 250 kOe and an electric
quadrupole interaction of 2.30 mm/sec with asymmetry parameter $\eta = 0.3$. The principle component of the efg is at right
angles to the magnetic hyperfine field. Johnson[20] also determined that the spins lay in the ac plane at an angle of $66.2°$ from
a axis, in agreement with results obtained by Narath[17] from
susceptibility and proton magnetic resonance measurements.

Kandel et al.[21] have studied the magnetic phases at
4.2 in external fields. A single crystal, grown from solution,
was oriented and cut so that the γ-ray propagation direction
and H_o were parallel to the easy axis a. The spectra in various
fields are shown in Fig. 15a, 15b, 15c and 15d. At $H_o = 0$
a four line spectrum is obtained, with a small absorption due
to non-magnetic $FeCl_2 \cdot 4H_2O$. For $H_o < H_1$ the spectrum
(Fig. 15b) consists of two superposed spectra of equal intensity corresponding to the external field H_o adding and sub-
tracting respectively from the hyperfine fields for the ions
in the spin down and spin up sublattice, respectively. For
$H_1 < H_o < H_2$, the spectrum (Fig. 15c) consists of two super-

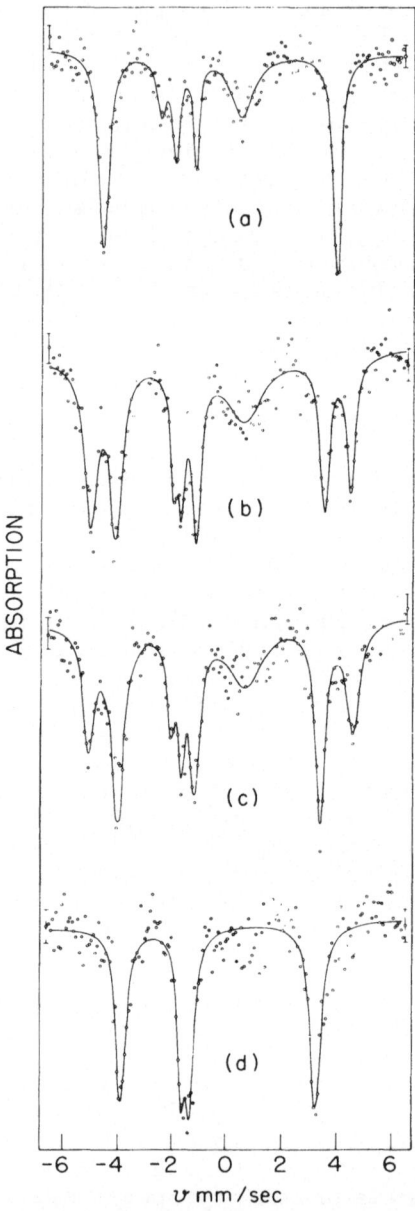

Fig. 15. Mössbauer spectra of $FeCl_2 \cdot 2H_2O$ at 4.2 K: a) H_o = (
b) H_o = 35 kOe; c) H_o = 42 kOe; d) H_o = 50 kOe. In all cases
$\gamma \parallel H_o \parallel$ easy axis (after Ref. 21).

posed spectra corresponding to the spin up and spin down sublattices, but now with relative intensities approximately 2:1. Since the majority spins have a smaller splitting than the minority spins, the sign of the hyperfine field is negative. For $H_2 < H_o$, a single spectrum is obtained (Fig. 15d). For all three phases, the sign magnitude and orientation of the efg is the same as in zero magnetic field, showing that the spins remain collinear in all three phases and that there is no spin canting. Moreover, the magnetic hyperfine interaction (exclusive of the applied field) is the same for all three phases (Fig. 16) indicating that the moment per ion is unchanged by increasing magnetic field or phase transitions.

The transition at 39 kOe is thus an AF to ferrimagnetic transition in which two spins are up and one spin down. The transition at $H_2 = 46$ kOe is a ferrimagnetic to P transition with all spins parallel. These results thus confirm the model proposed by Narath. [17]

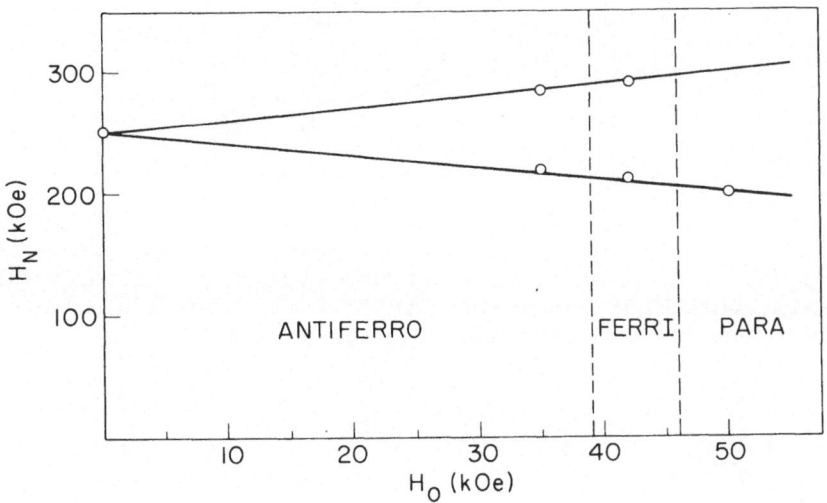

Fig. 16. H_n plotted as a function of H_o for the three phases of $FeCl_2 \cdot 2H_2O$ as determined from the spectra in Fig. 15.

FeCl$_2$ and FeBr$_2$. Simkin[22] has reported the Mössbauer spectroscopy of the metamagnetic transition in FeCl$_2$ and FeBr$_2$. In both cases, the spins are aligned antiferromagnetically along a threefold axis below the respective Néel temperatures (24 K for FeCl$_2$ and 11 K for FeBr$_2$). A metamagnetic transition is induced by an external magnetic field applied parallel to the threefold axis, at 10.5 kOe for FeCl$_2$ and 31.5 kOe for FeBr$_2$. In FeCl$_2$, the magnitude of the hyperfine field and the transition field are so low that the two sublattices cannot be resolved. In FeBr$_2$ marked changes in the FeBr$_2$ spectrum were observed on passage through the metamagnetic transition. The hyperfine field in FeBr$_2$ was found to be of positive sign (see Fig. 17) and a small change was observed in H_{hf} in going across the phase boundary. This change was shown by Simkin to be a change in the interionic dipole field in going from the AF to the P phase.

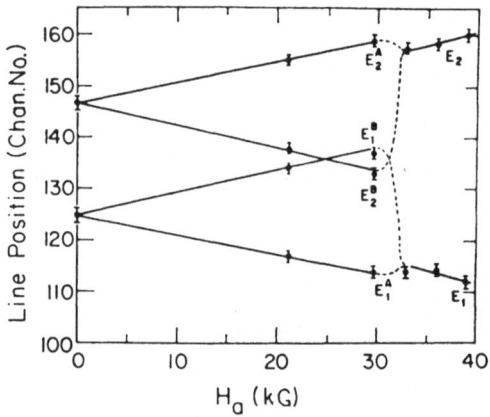

Fig. 17. Line position as a function of H_0 for single crystal FeBr$_2$, with H_0 ‖ easy axis. At the transition field the two sublattices become co-parallel (after Ref. 22).

FeCO$_3$. Forester and Koon[23] have made measurements in single crystals of FeCO$_3$. In this material the spins in the AF phase align antiparallel along the trigonal axis, below T_N = 38 K. An applied field $H_0 \approx 200$ kOe is necessary to induce the metamagnetic transition. As Forester and Koon applied fields up to 120 kOe their measurements concerned

the AF phase. They observed that the spectral lines for one
sublattice broadened with increasing T in a field H_o = 100 kOe,
while the other did not. They argued that the sublattice with
spins antiparallel to H_o will have a higher relaxation frequency
because those spins are more easily flipped in an external
field. A higher relaxation frequency means a greater line
width at a given temperature. They were thus able to determine that the sign of the hyperfine field in $FeCO_3$ is positive.

$FeCl_3$. Anhydrous $FeCl_3$ has been thoroughly studied by
Stampfel et al.[24] The Fe^{3+} are located in the interstices of a
hexagonal close-packed lattice of Cl ions, and below T_N = 8.7 K,
the spins order with a complex spiral-spin structure. Just
below T_N, the lines are broad, indicating spin relaxation effects.
Measurements of H_{hf} versus T fitted to Eq. (11) suggest that
the dimensionality of ordering is two (β = 0.156), but neutron
measurements show three dimensional order. Magnetic fields
parallel to the c-axis induce phase transitions. In fields
H_o < 15 kOe, a distribution of magnetic fields at the nuclei is
observed as expected from the spiral magnetic structure at
H_o = 0 determined by neutron diffraction. For 15 < H_o < 40 kOe,
a two sublattice model satisfactorily accountes for the spectra.
At H_o = 40 kOe, the spins changed orientation and the intensity
of the Δm = 0 lines increased suddenly indicating a transition
to an SF-like phase. The entire AF-SF phase boundary was
determined and H_{SF} was found to increase slightly with increasing temperature, up to the intersection with the AF-P
phase boundary. The possibility of a tri-critical point in the
$FeCl_3$ phase diagram is also discussed as an explanation for
the anomalously small value of β.

$FeCl_3 \cdot 6H_2O$. Carroll and Kaplan[25] have observed the
AF to P transition in magnetic fields in $FeCl_3 \cdot 6H_2O$, which
is antiferromagnetic below the Néel temperature T_N = 1.46 K.
In their experiment they held the temperature constant at 1.16 K
and increased the magnetic field, and obtained the spectra
shown in Fig. 18, which beautifully illustrates the AF to P
transition in this material at $H_{AFP} \approx 10$ kOe. The results
can be understood in terms of the MFA and Fig. 3. For
0 < H_o < H_{AFP}, there are two superposed spectra corresponding to spin up and spin down, with different field depen-
dences of the hyperfine field H_{hf} apart from the simple sub-
traction and addition of H_o to H_{hf} on the respective spin up
and spin down sublattices. For H_o > H_{AFP}, there is a single
spectrum in which H_{hf} varies roughly as Eq. (15). For fields
up to 20 kOe applied perpendicular to the spin axis they ob-

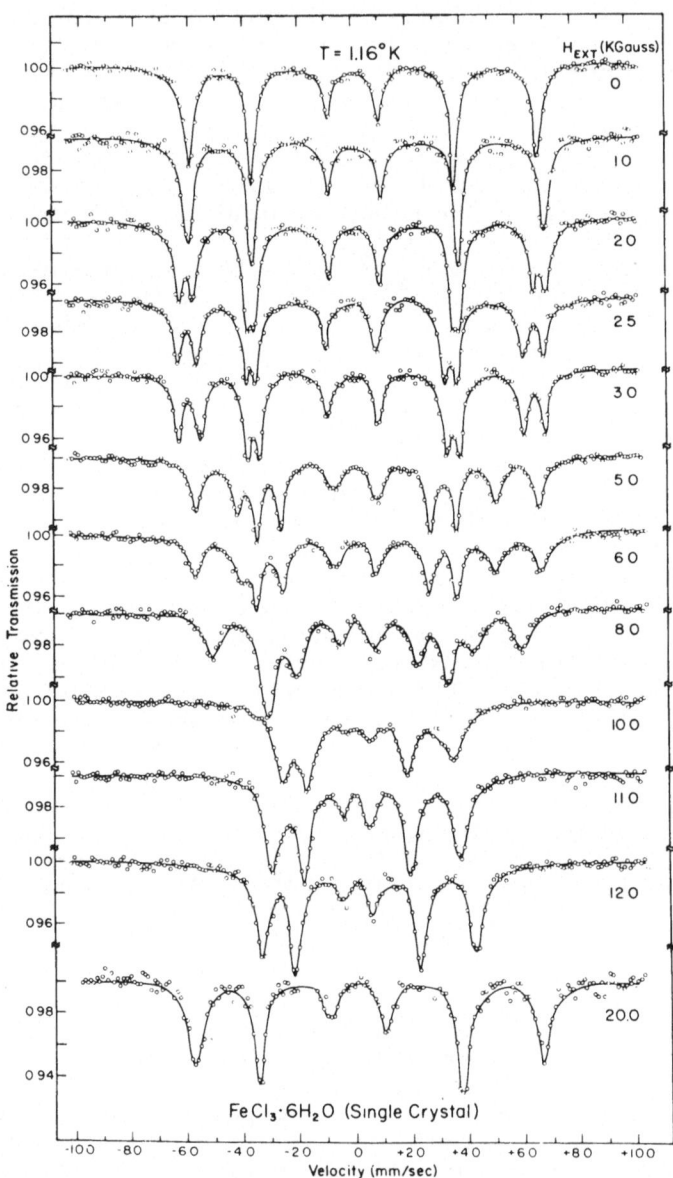

Fig. 18.   Mössbauer spectra of single crystal $FeCl_3 \cdot 6H_2O$ at 1.16 K, $\gamma \perp H_o$.   For the particular field direction, the AF to P transition occurs $8 < H_o < 10$ kOe.

served no transition. This is explicable in terms of the MFA because the AF to P transition with $H_o \perp a$ takes place at higher fields than for $H_o \parallel a$ (Eq. (9)). This is well illustrated experimentally by Shapira's ultrasonic attenuation study of the AF-P phase boundaries in FeF_2.[26]

V. CONCLUSION

We hope we have demonstrated that Mössbauer spectroscopy can be a very fruitful technique for studying magnetic field induced phases and phase changes in antiferromagnets. Of course, one needs external magnetic fields and single crystals, but the former are becoming increasingly available in the form of superconducting magnets and the latter can often be obtained if there is interest (and money). The study of the sublattice magnetization crossing the AF-P phase boundary can be made in any antiferromagnet containing iron and other Mössbauer nuclei and could prove interesting in the case of, for example, lower dimensional structures.

ACKNOWLEDGEMENTS

The participation of N. A. Blum, C. R. Abeledo, M. Weber, A. Misetich and L. Kandel in many of the experiments reported here is gratefully acknowledged. Y. Shapira, S. Foner and B. B. Schwartz provided useful and stimulating discussions and insights. We also thank Grace M. Lynch for her editorial assistance.

REFERENCES

+   Supported by the National Science Foundation.

1.   S. Foner in _Magnetism,_ Vol. I, edited by G. Rado and H. Suhl, Academic Press, New York, 1963, p. 383.

2.   Y. Shapira and S. Foner, Phys. Rev. B1, 3083 (1970).

3.   F. Keffer, _Handbuch der Physik,_ Vol. 18-2, Springer Verlag, Berlin. 1966.

4.   P. Heller, Phys. Rev. 146, 403 (1966).

5.   Y. Shapira, J. Appl. Phys. 42, 1588 (1971).

6. G.K. Wertheim, H.J. Guggenheim and D.N.E. Buchana;
    Phys. Rev. Lett. 20, 1158 (1968); J. Chem. Phys. 51,
    1931 (1969).

7. M.A. Lowe, A. Misetich and C.R. Abeledo, J. Phys.
    (Paris) 32, 1068 (1971).

8. C.R. Abeledo, R.B. Frankel, A. Misetich and N.A.
    Blum, J. Appl. Phys. 42, 1723 (1971).

9. C.R. Abeledo, R.B. Frankel, M.A. Weber and A.
    Misetich, AIP Conference Proceeding 10, 1168 (1973).
    A more complete version is being prepared for publi-
    cation in the Physical Review.

10. I.S. Jacobs, R.A. Beyerlein, S. Foner and J.P.
    Remeika, Int. J. Mag. 1, 197 (1971).

11. Y. Shapira, Phys. Rev. 184, 589 (1969).

12. N. Blum, A.J. Freeman, J.W. Shaner and L. Grodzins,
    J. Appl. Phys. 36, 1169 (1965).

13. F. van der Woude, Phys. Status Solidi 17, 417 (1966).

14. S. Foner and Y. Shapira, Phys. Lett. A29, 276 (1969).

15. N. A. Blum and R.B. Frankel, Bull. Am. Phys. Soc.
    12, 23 (1967) and unpublished data.

16. D.J. Simkin and R.A. Bernheim, Phys. Rev. 153,
    621 (1967).

17. A. Narath, Phys. Rev. A139, 122 (1965).

18. M.A. Lowe, C.R. Abeledo and A. Misetich, AIP
    Conference Proceeding 5, 307 (1972).

19. S. Chandra and G.R. Hoy, Phys. Lett. 22, 254 (1966).

20. C.E. Johnson, Proc. Phys.Soc. Lond. 88, 943 (1966).

21. L. Kandel, M.A. Weber, R.B. Frankel and C.R.
    Abeledo, Phys. Lett. (to be published).

22. D.J. Simkin, Phys. Rev. 177, 1008 (1969).

23. D.W. Forester and N.C. Koon, J. Appl. Phys. 40,
    1316 (1969).

24. J. P. Stampfel, W.T. Oosterhuis, B. Window and
    F. deS. Barros, Phys. Rev. B8, 4371 (1973).

25. T.X. Carroll and M. Kaplan, Chem. Phys. Lett.
    22, 564 (1973).

26. Y. Shapira, Phys. Rev. B2, 2725 (1970).

PHASE DETERMINATION BY KOSSEL CONE ANALYSIS UTILIZING THE
MOSSBAUER EFFECT*

J. P. Hannon and G. T. Trammell

Physics Department, Rice University

Houston, Texas 77001

ABSTRACT

Kossel line analysis offers a direct method for determining the phase of the X-ray structure factor of (molecular)
crystals. Phase determination is possible because the Kossel
lines result from the interference between the direct wave
emitted in a given direction with that which is Bragg reflected into the same direction. We give a theoretical discussion
of the problem.

I.   INTRODUCTION

Structure determination of complex molecules using X-ray
diffraction techniques is complicated by the fact that the
reflected intensity only gives the magnitude of the unit cell
scattering factor and not the phase. For this reason considerable effort has been devoted to devising methods for obtaining
the phase.

Typically one utilizes the interference between scattered
waves to determine the phase of the unit cell scattering amplitude relative to the phase of the wave scattered from a
limited number of impurity sites. For example in Mossbauer
diffraction one may substitute resonant Mossbauer
atoms into the sample to be studied. Because of the exceed-

*This work was supported by the National Science Foundation

181

ingly sharp resonance it is then possible to vary the phase
and amplitude of the resonantly scattered wave by Doppler
shifting with negligible variation in the nonresonant elect-
ronically scattered wave. Consequently the interference
between the waves is easily varied in a controlled manner.
As discussed by Black[1], and as demonstrated by Parak et
al.[2], it is then possible to extract the phase of the chemical structure factor of the unit cell relative to the structure factor due to the resonant scattering at the Mossbauer
sites. The problem is then reduced to determining the relative positions of the Mossbauer atoms within the unit cell.
In practice of course these experiments are very difficult,
and as discussed by Parak et al.[1] there are formidable problems in applying this procedure to the analysis of significant
biological molecules.

We propose a new method for phase determination by Kossel
line analysis. The Kossel lines occur when the emitting atoms
are located in the crystal[3-7]. The emitted photon is diffracted by the surrounding lattice, and the radiation pattern
outside the crystal exhibits a set of light and dark Kossel
cones, corresponding to the various internal Bragg reflections. The theory of the Kossel effect for X-rays and its
applications to precision determinations of crystal symmetries
and lattice parameters have been discussed extensively in the
literature[3-7]. However no one, to our knowledge, has suggested what could be the most important application of Kossel
line experiments: the determination of the phase of the X-ray
structure factor of (molecular) crystals.

The important feature is that the interference of the
direct wave emitted in a given direction with that Bragg
reflected into this direction (Fig. 1) allows the phase of the
latter to be obtained from the intensity distribution within
the Kossel line, as in the holographic technique where phase
information is obtained by "beating" the diffracted wave with
a primary reference wave.

                        II.   THEORY

Consider, for concreteness, a thick molecular crystal,
with isotropic Ml sources (radioactive isotopes, or e.g.
resonant Mossbauer atoms whose incoherent scattering when
illuminated constitute sources) substituted into the unit
cells near the top surface of the crystal (within the total

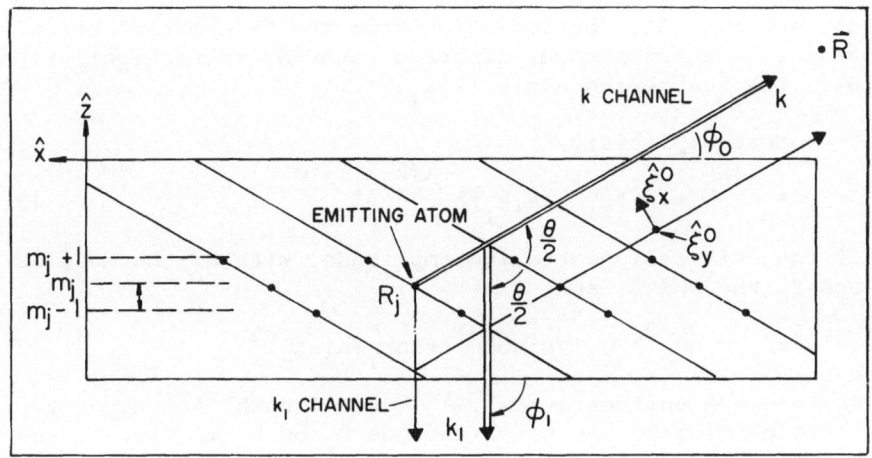

Fig. 1 Schematic of Bragg-Kossel emission. The wave emitted directly in the k-direction is indicated by the double line, and the wave reflected into the k-direction from the k_1-direction, which interferes with the direct wave, is indicated by the single line.

reflection extinction depth). At first we consider only one emitting site per unit cell at $\rho' = 0$. If the scattering is primarily Rayleigh scattering from the atomic electrons, then the intensity of the $\hat{\epsilon}_\lambda$ polarized component near a Bragg Kossel cone relative to the off-Bragg (background) intensity is[8]

$$
I_\lambda(\Delta\varphi)/I_{\lambda OB} = 1 - c^{-1}\cos\theta(\sin\varphi_o/\sin\varphi_1)^{\frac{1}{2}}[e^{i\phi_h}\{\Delta\varphi'/w_{\lambda h}
$$
$$
+ [(\Delta\varphi'/w_{\lambda h})^2 - 1]^{\frac{1}{2}}\}^{-1} + c.c.]
$$
$$
+ (\sin\varphi_o/\sin\varphi_1)|\Delta\varphi'/w_{\lambda h} + [(\Delta\varphi'/w_{\lambda h})^2 - 1]^{\frac{1}{2}}|^{-2}
$$

(1)

In (1) ϕ_h is the phase of the structure factor for the Bragg reflection "h"; $w_{\lambda h}$ is the angular width of the total reflection region, the thickness of the Kossel cone, for this reflection; $\Delta\varphi'$ is the difference of the cone angle of $\hat{k}$ and that of the center of the Kossel cone taken in the sense that $\Delta\varphi'>0$ for directions lying within the cone (as is well known $\Delta\varphi'$

differs from $\Delta\varphi$, the deviation from the "kinematic" Bragg angle, by a few seconds of arc because of refraction); finally θ is the scattering angle, $<(\hat{k}_1, \hat{k})$.

More explicitly

$$A_h e^{i\emptyset_h} = \sum_\rho f_{\lambda\lambda_1}^{(\rho)} (\underset{\sim}{k}, \underset{\sim}{k}_1) e^{-i(\underset{\sim}{k}-\underset{\sim}{k}_1)\cdot \underset{\sim}{\rho}} \tag{2}$$

is the unit cell scattering amplitude, with A_h the modulus and $\emptyset_h$ the phase, and

$$w_{\lambda h} = n\lambda_o^2 A_h C_\lambda (\pi \sin\theta)^{-1} (\sin\varphi_1/\sin\varphi_o)^{\frac{1}{2}} \tag{3}$$

where n = # unit cells/cm^3, λ_o = wavelength; λ = x, or y, corresponding to the polarizations $\hat{\varepsilon}_x$ or $\hat{\varepsilon}_y$ in Fig. 1, and $C_x = \cos\theta$, $C_y = 1$, and φ_1 and φ_o are the angles $\hat{k}_1$ and $\hat{k}$ make with the crystal surface (Fig. 1). We also have $\Delta\varphi' = \Delta\varphi + n\lambda_o^2 A_o (2\pi\sin\theta)^{-1} (\sin\varphi_o/\sin\varphi_1 + 1)$ where the second term represents the deviation of the center of the total reflection region from the kinematic Bragg angle, with A_o the forward scattering structure factor ($\hat{k} = \hat{k}_1$ in 2). A_o has a small positive imaginary part.

In Eq. (1) the first term, 1, represents the intensity of the wave emitted directly into the $\hat{k}$ direction, the last term that Bragg reflected into this direction, and the second term the interference of these two waves.

For $|\Delta\varphi'| < w_{\lambda h}$ the reflected wave has modulus $(\sin\varphi_o/\sin\varphi_1)^{\frac{1}{2}}$ [hence total reflection since waves emitted in the solid angle $d\Omega_1$ are Bragg reflected into solid angle $d\Omega_8) = (\sin\varphi_o/\sin\varphi_1)d\Omega_1$], and its phase is fixed by the stipulation that $\text{Im}[(\Delta\varphi'/w_{\lambda h})^2 - 1]^{\frac{1}{2}} > 0$. The phase of the reflected wave amplitude

$$r_\lambda(\Delta\varphi) \equiv -e^{i\emptyset_h} [\eta' + (\eta'^2 - 1)^{\frac{1}{2}}]^{-1} (\sin\varphi_o/\sin\varphi_1)^{\frac{1}{2}}, \tag{4}$$

with $\eta' \equiv \Delta\varphi'/w_{\lambda h}$ is thus $\emptyset_h$ for $\eta' < -1$ (outside the Kossel cone), $\emptyset_h + \pi$ for $\eta' > +1$ (inside the Kossel cone), and $\emptyset_h + \pi - \tan^{-1}[(1-\eta'^2)^{\frac{1}{2}}/\eta']$ for $|\eta'| < 1$ (within the width of the Kossel line). The phase of r_λ increases by π as η' goes from -1 to $+1$, and it is this that allows $\emptyset_h$ to be inferred from intensity measurements.

If $\cos\theta$ is small and $\varphi_o = \varphi_1$ (Bragg planes parallel to the surface) then 1) becomes

$$I_\lambda(\Delta\varphi)/I_{\lambda OB} = \left|1-e^{i\emptyset_h}[\eta'+(\eta'^2-1)^{\frac{1}{2}}]^{-1}\right|^2 \tag{5}$$

If then $0 < \emptyset_h < \pi$, $I_\lambda(\Delta\varphi')/I_{\lambda OB}$ will go to zero at some point in the total reflection region $|\eta'| < 1$, whereas if $\pi < \emptyset_h < 2\pi$ it will become 4 at some point in this region.

In the general case we have for $|\eta'| > 1$

$$I(\Delta\varphi')/I_{OB} = 1-2C_\lambda^{-1}\cos\theta(\sin\varphi_o/\sin\varphi_1)^{\frac{1}{2}}[\eta'+(\eta'^2-1)^{\frac{1}{2}}]^{-1}\cos\emptyset_h$$

$$+ (\sin\varphi_o/\sin\varphi_1)[\eta'+(\eta'^2-1)^{\frac{1}{2}}]^{-2} , \tag{6}$$

where the radical is positive if $\eta' > 1$ and negative if $\eta' < -1$; and for $|\eta'| < 1$

$$I(\Delta\varphi')/I_{OB} = 1-2C_\lambda^{-1}\cos\theta(\sin\varphi_o/\sin\varphi_1)^{\frac{1}{2}}[\eta'\cos\emptyset_h$$

$$+ (1-\eta'^2)^{\frac{1}{2}}\sin\emptyset_h] + \sin\varphi_o/\sin\varphi_1 \tag{6'}$$

Comparing (6) and (6') we see that the antisymmetric term (about $\eta' = 0$) is proportional to $\cos\emptyset_h$, whereas the symmetric interference term is proportional to $\sin\emptyset_h$.

III.   PHASE DETERMINATION

In Fig. 2 we plot $I(\Delta\varphi')/I_{OB}$ (with $\cos\theta \doteq 1$, $\varphi_1 = \varphi_o$) as a function of η', for several values of $\emptyset_h$. The shape of the line is clearly a very sensitive function of $\emptyset_h$, and the line usually contains a bright ring (intensity maximum) and a dark ring relative to background. In a centrosymmetric crystal, with the emitting atom at a center of symmetry, $\emptyset_h = 0$ or π. Then if $\theta < \pi/2$ the Kossel cone will have an inner dark ring and an outer bright ring if $\emptyset_h = 0$ and the opposite if $\emptyset_h = \pi$. If $\theta > \pi/2$ there is a reversal of order because of the $\cos\theta$ dependence of the interference term. Thus $\emptyset_h$ can be obtained for all the reflections of a centrosymmetric molecule by qualitative examination of the Kossel cones recorded on a photographic plate.

It is apparent from (6), (6') or Fig. 2 that high reso-lution measurements of the Kossel cone structure would deter-mine A_h and $\emptyset_h$. Unfortunately the widths of these lines, w_h (Eq. 3), are $\sim$ few seconds of arc. To obtain the Kossel lines with a resolution $\sim$ line width would require that the detector be located at a distance $\sim 10^5$ x linear dimension of the

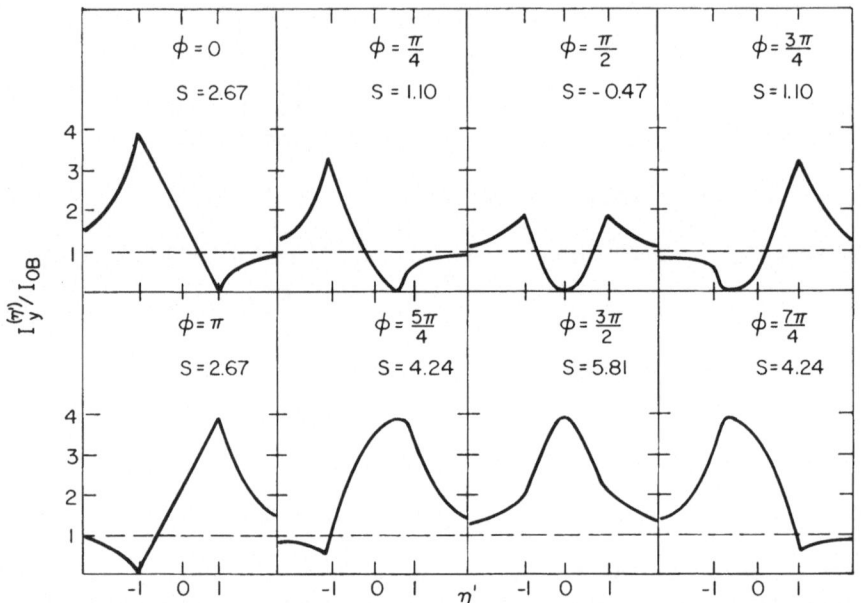

Fig. 2 Rocking curves for the Kossel line intensity (relative
to the off Bragg intensity) for a symmetric Bragg reflection
channel for different values of the phase angle Ø of the unit
cell scattering amplitude. The region $|\eta'| < 1$ corresponds
to the Bragg total reflection region. The values of S give
the integrated intensities (relative to off Bragg) for the
various curves.

crystal. Practically, A_h should be determined by X-ray diffraction techniques, then $\emptyset_h$ can be determined by measurements
using "poor" resolution, say 10-50 sec of arc.

    Consider the integrated relative brightness excess

$$S'_{2\Delta} = \sum_\lambda \int_{-\Delta}^{\Delta} [I_\lambda(\Delta\varphi') - I_{\lambda OB}] d(\Delta\varphi') / \sum_\lambda \int_{-\Delta}^{\Delta} I_{\lambda OB} d(\Delta\varphi')$$

$$= (4\Delta)^{-1} \sum_\lambda \int_{-\Delta}^{\Delta} [I_\lambda(\Delta\varphi')/I_{\lambda OB} - 1] d(\Delta\varphi') \qquad (7)$$

$$\doteq \frac{w_h}{2\Delta} \frac{\sin\varphi_o}{\sin\varphi_1} \{(4/3)(1+\cos\theta) - \pi\cos\theta(\sin\varphi_1/\sin\varphi_o)^{\frac{1}{2}} \sin\emptyset_h\} ,$$

$$\Delta \gg w_h .$$

where $w_h = w_{\bar{y}h}$ is given by (3) with $\lambda = y$. If A_h is known then measurement of the integrated intensity in an angular region 2Δ centered on the Kossel line will determine $\sin\emptyset_h$ through (7). The correct quadrant (first or second, third or fourth) can then be determined by qualitative observation of the dark-bright ring order (Fig. 2) as in the case of the centrosymmetric crystal.

As an example $2w_h \approx 2.6$ sec of arc for the (020) Bragg reflection in $K_3Fe(CN)_6$[2). If we take $2\Delta = 10$ sec then (7) gives $S'_{2\Delta} = 0.33$, an excess counting rate over that of background of 33 counts per hundred. In fact $\emptyset_h$ is zero for this reflection, but if instead $\emptyset_h$ were, e.g. $\pi/2$ or $3\pi/2$ then one would measure excess relative intensities $S'_{2\Delta} = -0.08$ or $+0.74$ respectively.

For the (800) reflection in myoglobin[9) $2w_h \approx .52$ sec of arc. Again taking $2\Delta = 10$ sec arc we get $S'_{2\Delta} = .07$. $\emptyset_h$ is also zero for this reflection. If it were $\pi/2$ or $3\pi/2$ then $S'_{2\Delta}$ would be $-.013$ or $+.14$ respectively.

Naturally this method of phase determination for large molecules will present difficult experimental problems. The intensities and contrast are low, and the lines are very narrow. On the other hand if the emitted radiation is recorded, e.g. on a photographic plate, one would obtain simultaneously the Kossel cones from all (or a large fraction of) the Bragg reflections. This does represent a straightforward method for phase determination by what is essentially a holographic technique[10), and it may be the best way of obtaining the phases for large molecules[11).

IV.   GENERALIZATIONS

In the above development we have assumed a single emitting site in each unit cell fixed at $\rho' = 0$. It is simple to take the zero point and thermal motion of the emitters into account and to allow as well for several emitting sites in the unit cells.

We first note that the Kossel cones of the doppler shifted (recoil) γ-rays will generally differ in angular position from those of the recoilers γ-rays by less than the width of the Kossel cone, w_h. Hence both the recoil and recoilless γ-rays contribute to the observed intensity. This comes from the fact that when a phonon is created or absorbed in the radiation process it will generally shift the energy of the γ-ray by less than about 0.1 ev. If the individual atoms are to be resolved then γ-ray wavelengths less than 1A or γ-ray energies greater than 10^4 ev are required. Thus $\delta k/k \lesssim 10^{-5}$, and since the shifts in positions of the bragg angles $\delta\varphi_B \approx \delta k/k$ we obtain $\delta\varphi_B \lesssim 10^{-5}$ rad $\approx$ 1 sec arc $\lesssim w_h$.

If now there are ν emitting sites in each unit cell the formulas we have obtained for the relative intensities must be modified according to the following substitutions

$$\cos\varphi_h \rightarrow \mu\cos(\varphi_h + \varphi_n)$$

$$\sin\varphi_h \rightarrow \mu\sin(\varphi_h + \varphi_n) \ , \tag{8}$$

where $\mu e^{i\varphi_n}$ is the structure factor for the emitting sites,

$$\mu e^{i\varphi_n} = \frac{1}{\nu}\sum_{\rho'}\langle e^{-i(\underset{\sim}{k}-\underset{\sim}{k}_1)\cdot\rho'}\rangle$$

$$= \frac{1}{\nu}\sum_{\rho'} e^{-\frac{1}{2}\langle[(\underset{\sim}{k}-\underset{\sim}{k}_1)\cdot(\rho'-R')]^2\rangle} \cdot e^{-i(\underset{\sim}{k}-\underset{\sim}{k}_1)\cdot\underset{\sim}{R}'} \tag{9}$$

where $R' = \langle\rho'\rangle$ are the equilibrium positions of the emitters and where in the second line of (9) we have used the Debye-Waller approximation.

If $\nu > 1$ it is necessary to first obtain the relative positions of the emitting sites together with their Debye-Waller factors so that μ and φ_n may be computed. Then with these values and our previous expressions modified in accordance with (8) φ_h can be determined by Kossel line analysis. If there are only a small number of emitters per unit cell then their positions should be ascertainable by Mossbauer and/or X-ray diffraction experiments.

If the sources are isotropic El or E2 emitters rather than M1, the only change in the previous results is in the interference term. Thus for El emitters (1) becomes

$$I_\lambda(\Delta\varphi)/I_{\lambda OB} = 1-C_\lambda(\sin\varphi_0/\sin\varphi_1)^{\frac{1}{2}}[e^{i\varphi_h}\{\Delta\varphi'/w_{\lambda h}$$

$$+ [(\Delta\varphi'/w_{\lambda h})^2-1]^{\frac{1}{2}}\}^{-1}+c.c.]+(\sin\varphi_0/\sin\varphi_1|\Delta\varphi'/w_{\lambda h}$$

$$+ [(\Delta\varphi'/w_{\lambda h})^2-1]^{\frac{1}{2}}|^{-2} \ , \tag{10}$$

and $S'_{2\Delta}$ (Eq. (7)) becomes

$$S'_{2\Delta} \doteq \frac{w_{\gamma h}}{2\Delta} \frac{\sin\varphi_o}{\sin\varphi_1} \left\{ 4/3(1+\cos\theta) - \frac{\pi}{2}(1+\cos^2\theta)\left(\frac{\sin\varphi_1}{\sin\varphi_o}\right)^{\frac{1}{2}} \sin\phi_h \right\}.$$

(11)

and in the case of isotropic E2 emitters

$$I_\lambda(\Delta\varphi/I_\lambda(OB)) = 1-C'_\lambda \cos\theta(\sin\varphi_o/\sin\varphi_1)^{\frac{1}{2}}[e^{i\phi_h}\{\Delta\varphi'/w_{\lambda h}$$
$$+ [(\Delta\varphi'/w_{\lambda h})^2-1]^{\frac{1}{2}\,-1}+c.c.]+\sin\varphi_o/\sin\varphi_1|\Delta\varphi'/w_{\lambda h}$$
$$+ [(\Delta\varphi'/w_{\lambda h})^2-1]^{\frac{1}{2}}|^{-2},$$

(12)

and

$$S'_{2\Delta} \doteq \frac{w_{\gamma h}}{2\Delta} \frac{\sin\varphi_o}{\sin\varphi_1} \left\{ 4/3(1+\cos\theta) - \pi\cos^3\theta\left(\frac{\sin\varphi_1}{\sin\varphi_o}\right)^{\frac{1}{2}} \sin\phi_h \right\}.$$

(13)

In (12) $C'_x = \cos 2\theta$ and $C'_y = 1$. Comparing these with our previous expressions (1) and (7) it is seen that the major features of the Kossel cones are the same for E1 and E2 as for M1 emitters.

REFERENCES

1. P. J. Black, Nature 206, 1223 (1965).

2. F. Parak, R. L. Mossbauer, U. Biebl, H. Formanek, W. Hoppe, Z. Phys. 244, 456 (1971).

3. W. Kossel, H. Loeck, and H. Voges, Z. Physik 94, 139 (1935).

4. Max von Laue, Rontgenstrahleninterferenzen (Akademische Verlagsgescellschaft, Frankfurt am Main, 1960) pp. 430-448).

5. R. W. James, The Optical Principles of the Diffraction of X Rays (Cornell University Press, Ithaca, N.Y., 1965) pp. 413-457.

6. See for example R. Tixier and C. Wache, J. Appl. Cryst. 3, 466 (1970).

7. J. P. Hannon and G. T. Trammell, Mossbauer Effect

Methodology, Vol. 8 (Plenum Publishing Co., New York, 1973) pp. 25-40.

8.   J. P. Hannon, N. J. Carron, and G. T. Trammell, accepted for publication in Physical Review.

9.   For a discussion of the crystalline properties of myo-globin see G. Bodo, H. M. Dintra, V. C. Kendrew, and H. W. Wyokoff, Proc. Roy. Soc. A253, 70 (1959); or F. Parak, Ph.D. thesis, Technische Hochschule Munchen (1970), unpublished.

10. The emitters must be located at well defined positions in the unit cell. If the mean positions of the emitters were spread around a given site in the unit cell by as much as a wavelength the phase information would be obliterated (the zero point, thermal, and even recoil motion of the emitter however will not obliterate the phase information so long as the quantum mechanical mean position of the emitter is fixed to within a small fraction of a wavelength).

11. If and when an X-ray laser is constructed it will be extremely difficult to use it to determine phases of crystal structure factors because in such a process as the crystal's orientation is changed to give the various Bragg reflections the path difference of the laser to the detector and the laser to crystal to detector would have to be controlled to within a wavelength. On the other hand a hologram of a single molecule (with atomic resolution) certainly seems precluded by the destruction of the molecule (J. R. Breedlove Jr. and G. T. Trammell, Science 170, 1310 (1970)).

RETRIEVAL, DOCUMENTATION, AND EVALUATION OF MÖSSBAUER SPECTROSCOPY DATA

John G. Stevens and Virginia E. Stevens

University of North Carolina at Asheville
Asheville, North Carolina 28804

INTRODUCTION

In the fifteen years since its observation was first reported, Mössbauer spectroscopy has developed into such a
sophisticated technique that it used routinely by many scientists today. Approximately seven thousand articles have
been published on Mössbauer spectroscopy with the current
annual rate approaching one thousand. The activity of the
Mössbauer Effect Data Index (MEDI) group has been the retrieval and documentation, and more recently, the evaluation
of the Mössbauer literature. The first group, consisting of
A. H. Muir, Jr., K. J. Ando, and H. M. Coogan, originated
at North American Aviation (now North American Rockwell).
Under their direction the North American Aviation Science
Center informally published and distributed on request the
Mössbauer Effect Data Index, Issue 1 (1958-1962), Issue 2
(1958-1963), and Issue 3 (1958-1964). Following these three
issues the Mössbauer Effect Data Index 1958-1965[1] was published and soon became known as the "Blue Book." The decrease in emphasis on basic science with its resultant cutback
in funding put the Index into the files until it was revived by
our group. With the advice and direction of Dr. Art Muir, Jr.,
the first annual MEDI covering the 1969 literature was published[2]. At this time there were no elaborate systems or
computer programs developed to handle the articles, so every

computer card was hand-sorted and the different listings and
the author file created manually. The following year, while
frantically attempting to "de-bug" programs and having to re-
punch thousands of cards unacceptable to the computer, we
longed for the peaceful days of hand-sorting. But now the prc
gramming is sophisticated and detailed and this year the last
vestige of hand-sorting will disappear. More of this aspect
of the operation is discussed under Documentation.

In the early months of our operation questionnaires were
distributed to some Mössbauer spectroscopists so we could
better estimate the value and chart the future of MEDI, as
well as establish the practice of encouraging input by the com
munity. The 98% response made the effort seem worthwhile,
so we continued. Input has also continued and runs the gamut
everything from one humorous suggestion we give gold stars
"a la Michelin" indicating the quality of the papers to slightly
agitated letters telling of overlooked articles to helpful ones
suggesting changes or bringing errors to our attention.

Financing the first MEDI (1969) was a real shoestring en-
deavor. Partial funding came from the North Carolina Board
of Science and Technology; the rest from a variety of places
who were willing to provide services. All Indices[3-6] since
then have received support from the National Standard Reference Data System, previous sales of MEDI's (1969 to the present), advertisements and other incidental revenues.

RETRIEVAL

The methods of searching for Mössbauer articles are
varied, falling into four catagories (1) computer searches,
(2) abstract services, (3) personal searches, and (4) miscel-
laneous retrieval resources. Table 1 lists all of those used
in searching the 1973 literature and provides an analysis of
their efficiency as of February 1, 1974. (This table is subjeci
to change upon completion of the 1973 search when all the statistics will be available.) Chemical Condensates, a service
of Chemical Abstracts, offers by far the best coverage.
ASCATOPICS is the most current, often listing references

Table 1

Summary of Retrieval Methods

Retrieval Methods	% of Total References Found	% of Total References Found Only By Listed Method
I. Computer Searches		
A. Chemical Condensates	79	8
B. ASCATOPICS (Institute for Scientific Information)	61	4
C. Nuclear Science Search	12	0.2
D. Scientific Documentation Corporation	35	0.2
II. Abstracts Services		
A. Physics Abstracts	41	0.2
B. Current Physics Titles	40	0.2
C. Biological Abstracts	1	0.0
III. Personal Searches		
A. Publishers' Lists	2	2
B. Search of 35 Journals	30	3
IV. Miscellaneous Retrieval Resources		
A. Sent without Request	10	5
B. Special Reference Lists Sent Periodically by Various Groups or Authors Writing Reviews	5-20	-

in journals before they are even on library shelves. Some of the other services and personal searches are much less impressive.

The second column of Table 1 represents the percentage of articles that are retrieved exclusively by each particular method. One can see that in our annual analysis of methods to be retained and those to be dropped, in some cases we are confronted with the difficult problems of whether or not to expend the time and money for a search which provides only one

percent additional references. The table also shows the significant number of articles found which otherwise would not
be by (1) checking scientific publishers' lists, (2) searching
35 pertinent journals personally, and (3) being sent reprints
by the authors themselves. Finally the table indicates that
a personal search, even the thorough one used by MEDI,
while it does uncover articles which might go undetected,
only gives thirty percent coverage. Since an individual scientist would probably not maintain such an extensive list, he
could not hope to find more than half of the articles relative
to his area — unless he gave up his research to search.

Reprints of all articles are obtained either from the
authors or from one of several libraries. Aiding the project
in acquisition of articles from the more obscure foreign journals have been Dr. Mikio Takano of Konan University and Prof
G. N. Belozerskii of the Leningrad State University. In some
instances they have also helped by abstracting pertinent data
from the Japanese or Russian articles. We also select a few
Russian articles each year, otherwise untranslated, for
translation.

On the whole, thorough searching becomes more and more
difficult to achieve. Despite the yearly improvement of our
retrieval techniques, Mössbauer spectroscopy has come into
its own as a technique and is, therefore, often not a central
topic of a paper.

DOCUMENTATION

All reference information and the pertinent Mössbauer
data are written onto special coding forms from which cards
can be punched. A unique format was developed to create
upper case and lower case letters in our printout. (Normally
only upper case letters are produced.) This, of course, permits much easier reading of the Index and avoids possible
misunderstandings in the data section.

Besides this program there are about thirty others used
to process the information, sort it, and print it in book format. There are also several internal checks made using

NI-61, 67.4 KEV TRANSITION

SOURCE	S-TEMP	ABSORBER	A-TEMP	SHIFT	QS	REMARKS	REF
Ni-V alloy	He	Ni-Pd alloys	He	--		fa, IS, & FI vs Ni conc	72T008
Cr-Ni alloy	4.2	Ni-Pt alloys	4.2			HI vs Ni concentration	72F015
Ni-V alloy		Ni2MnGa	4.2			HA=125 kOe	72L024
Ni-V alloy		Ni2MnIn	4.2			HA=141 kOe	72L024
Ni-V alloy		Ni2MnSb	4.2			HA=47 kOe	72L024
Ni-V alloy		Ni2MnSn	4.2			HA=125 kOe	72L024

CODE	TOPIC	REFERENCE
72F015	NI-61	W A Ferrando, R Segnan, and A I Schindler, Phys Rev B 5, 4657-64(1972), Matrix and Impurity-cluster Polarization in Ni-Pt and Ni-Pd Alloys
72L024	NI-61	J C Love and F E Obenshain, Hyperfine Fields at Ni Sites in Heusler Alloys, in "AIP Conf Proc-No 5, Magnetism and Magnetic Materials-1971" (Seventeenth Annual Conference, Chicago, 1971), edited by C D Graham, Jr and J J Rhyne (American Institute of Physics, New York, 1972), part 1, p 538
72T008	NI-61	J E Tansil, F E Obenshain, and G Czjzek, Phys Rev B 6, 2756-808(1972), Nb1 Mossbauer Effect in Ni-Pd Alloys

Figure 1. Data/Reference page from the Mössbauer Effect Data Index-Covering the 1972 Literature.

the computer. But since the computer can only determine
format errors and we get used to seeing our own mistakes,
print-outs are sent to two or three Mössbauer spectroscopists
who go over them for errors.

Along with every data entry (see upper portion of Figure 1)
and reference entry (see lower portion of Figure 1) are included, but not printed out, a list of keywords. Up to twelve
keywords can be included for each data entry and nineteen for
each reference entry. Table 2 lists some of the keywords
that are used. With these keywords we have capabilities of
searching our reference and data files. The program, written
in Assembler Language to minimize computer time, can perform one hundred or more simultaneous searches.

There are now one hundred Mössbauer transitions that
have been reported in the literature (see Table 3). Keeping
up with the nuclear parameters of these transitions is a major
task in itself. We have been assisted, both directly and indirectly, by the Nuclear Data Project at Oak Ridge National
Laboratory. Tabulated nuclear parameters (nuclear radii,

Table 2

Sample of the List of Keywords Used for Reference and
Data Listings in the Mössbauer Spectroscopy Data Bank

actinides	bonding information
alloys	Bragg scattering
alpha decay-radiation damage	Brownian movement
angular correlation	catalyst experiment
antiferromagnetic substances	chemical after-effect
apparatus	chemical identification
area under absorber line	chemical kinetics
assymetry parameter	colloids
backscattering detector	computer program
backscattering experiment	conductivity
biological compounds	corrosion

Note: Any element (e.g. Fe, Sn, Eu, Xe...), functional
group (e.g. methyl, phenyl, propyl, $SO_4^=$, $PO_4^=$...) or
language may be used as a keyword.

Table 3

Mössbauer Periodic Table

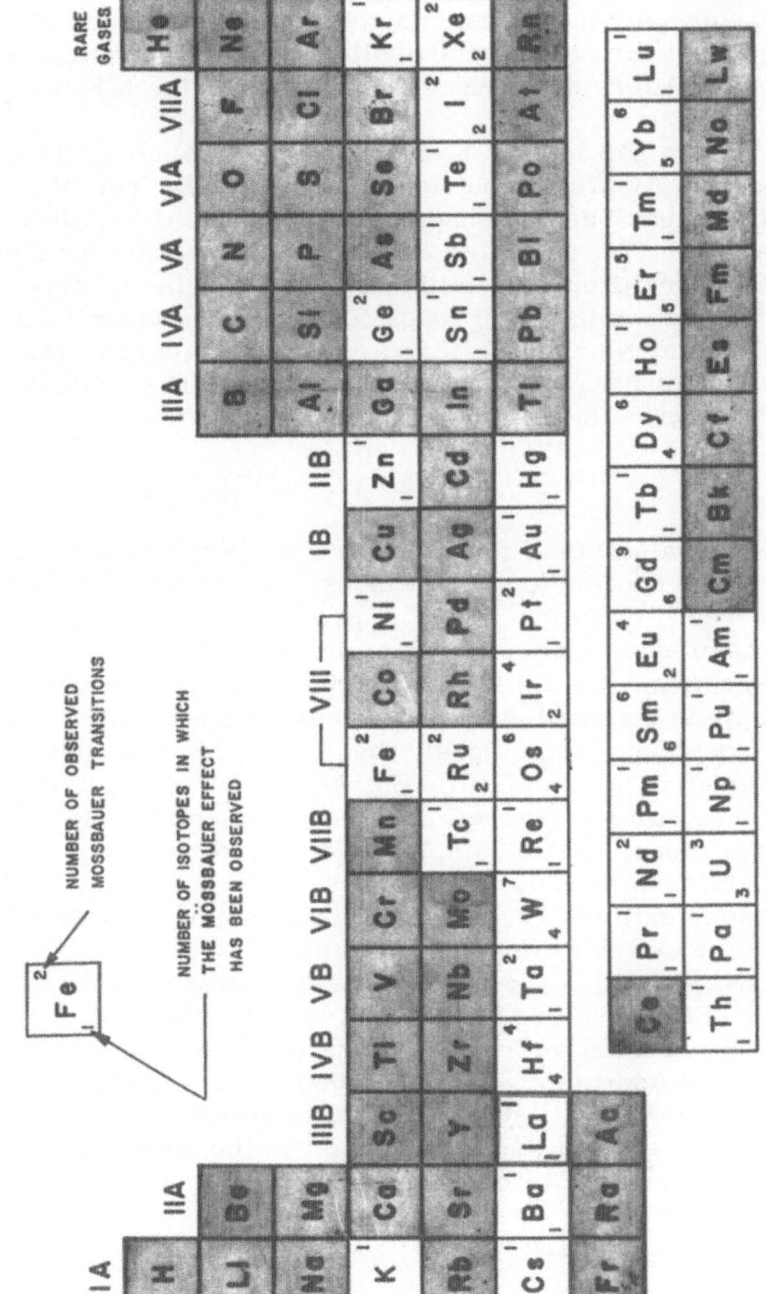

quadrupole moments, magnetic moments, and half lives) from
Mössbauer studies are published for all transitions reported
in the literature covered by the particular MEDI.

The gap left by the last MEDI published by the North American Aviation group and the first MEDI from the present
group has been a conspicuous one since it includes three
years (1966-1968) and one-fourth of the total number of Mössbauer references. With the support of the National Science
Foundation the two groups (Drs. Art Muir, Jr. and Ronald
Grant at North American Rockwell and ourselves) have been
collaborating to complete the series. The "missing years" of
MEDI should be available by the end of the year.

EVALUATION

Evaluation of the literature is always an area which requires caution for two reasons: (1) it is impossible for one
person, or even a group of people, should one be formed, to
make expert judgment on articles covering such a broad
spectrum; (2) the tenacity and emotion with which most scientists defend their papers would make the MEDI group very
unpopular. So obviously the "gold star" idea previously mentioned is not realistic. But we are looking to the Mössbauer
community for direction and suggestions in this area. Does
the group remain completely objective and confine itself to
publishing MEDI, which by its nature will include all Mössbauer articles regardless of their scientific worth? Or should
more critical evaluations be made?

Efforts so far have resulted in a number of reviews
written for practical use being included in MEDI. The most
recent addition to MEDI has been a series of papers in which
each concerns itself with a particular isotope. Each of these
is written by a leading scientist in the specific area covered
by the review. The emphasis is on completeness and a description of the "state of the art." Table 4 summarizes the
reviews in MEDI.

Indirectly we also aid in the evaulation of the literature by
others by providing them with lists of pertinent references or

Table 4

Reviews and Useful Papers in the Mössbauer Effect Data Index

Title	Author	Reference
Suggested Nomenclature and Conventions	R. H. Herber	2
Useful Information for ^{57}Fe Mössbauer Spectroscopy	J. G. Stevens R. S. Preston	3
An Introduction to Electric Quadrupole Interactions in Mössbauer Spectroscopy	B. D. Dunlap	3
Calibration in Mössbauer Spectroscopy	P. A. Flinn	4
Nomenclature and Conventions for Reporting Mössbauer Spectroscopic Data	J. J. Zuckerman et al.	4
Nuclear Gamma Resonance with ^{61}Ni	F. E. Obenshain	5
^{237}Np Mössbauer Spectroscopy	J. A. Stone	5
A Review of the Mössbauer Spectroscopy of ^{99}Ru and ^{101}Ru	M. L. Good	5
^{121}Sb Mössbauer Spectroscopy	L. H. Bowen	5
^{197}Au Mössbauer Spectroscopy	L. D. Roberts	6
^{151}Eu Mössbauer Spectroscopy	N. N. Greenwood	6
^{129}I Mössbauer Spectroscopy	H. de Waard	6

obtaining articles that are hard to acquire. In addition, some of our own efforts, not directly related to MEDI but certainly a spin-off, have been the publication of four reviews[7-10] and a critical compilation of nuclear moments from Mössbauer spectroscopy.[11]

CONCLUSION

While collecting and organizing the Mössbauer literature we have observed two interesting trends. The first is the continued preponderance of work centered around the 14.41 keV ^{57}Fe and the 23.87 keV ^{119}Sn transitions. Although there are

98 other Mössbauer transitions, work on all these comprises less than 15% of the literature. The more astonishing fact is that this percentage decreases annually. It is true that a number of transitions are difficult to observe, but there are still a reasonable amount which are quite practical for use. What is needed are more scientists who are willing to branch out and study them.

Our second observation is the growing amount of research being done in the Eastern European countries. One dramatic evidence of this increase was the large attendance and number of papers presented (186) at the 5th International Conference on Mössbauer Spectroscopy held in Bratislava, Czechoslovakia in 1973.

All in all, it is quite apparent Mössbauer spectroscopy is no longer confined mainly to large, well-financed laboratories and universities, but has become a familiar and useful technique used in smaller laboratories and developing countries as well. One problem now is to help these one-man or small group operations to maintain contact with the community at large. Since in their circumstances library holdings are often sparse and isolation a way of life, the lack of communication can produce distorted science. The community as a whole needs to address itself to this problem. Here, again is an area in which the MEDI group may be able to perform a more valuable service. And again, as we remember past critcisms and suggestions which have led to improvement in MEDI, we recognize the importance of continued input by the Mössbauer community.

REFERENCES

1.   A. H. Muir, Jr., K. J. Ando, and H. M. Coogan, Mössbauer Effect Data Index 1958-1965, (Interscience Publishers, New York, 1966).
2.   J. G. Stevens and V. E. Stevens, Mössbauer Effect Data Index - Covering the 1969 Literature (IFI/Plenum, New York, 1970).

3.   J. G. Stevens and V. E. Stevens, Mössbauer Effect Data Index-Covering the 1970 Literature (IFI/Plenum, New York, 1972).

4.   J. G. Stevens and V. E. Stevens, Mössbauer Effect Data Index-Covering the 1971 Literature (IFI/Plenum, New York, 1972).

5.   J. G. Stevens and V. E. Stevens, Mössbauer Effect Data Index-Covering the 1972 Literature (IFI/Plenum, New York, 1973).

6.   J. G. Stevens and V. E. Stevens, Mössbauer Effect Data Index-Covering the 1973 Literature (IFI/Plenum, New York, 1974).

7.   J. G. Stevens, J. C. Travis, and J. R. DeVoe, Anal. Chem. 44, 384R (1972).

8.   J. G. Stevens, Magn. Res. Rev., 2, 97 (1973).

9.   J. G. Stevens and L. H. Bowen, Anal. Chem., 46, xx (1974).

10.   J. G. Stevens, Magn. Res. Rev., 3, xx (1974).

11.   B. D. Dunlap and J. G. Stevens (submitted for publication).

METHODOLOGY

COMPARISON OF TECHNIQUES FOR FOLDING AND UNFOLDING MÖSSBAUER SPECTRA FOR DATA ANALYSIS

Tsung-Ming Lin and Richard S. Preston

Department of Physics
Northern Illinois University
DeKalb, Illinois 60115

A Mössbauer spectrum of intensity I vs Doppler energy shift δ can be represented by

$$I(\delta) = C\int_{-\infty}^{\infty} S(\delta-E)\exp[-t\cdot\sigma(E)]dE + B \qquad (1)$$

where S is the source line shape, σ is the cross section for nuclear absorption of recoilless gamma rays, and t represents the absorber thickness including the effects of the Mössbauer fraction for the absorber. The constant C includes the magnitude of the flux incident on the absorber and the Mössbauer fraction of the source. The constant B is that part of the radiation transmitted by the absorber which could not have been resonantly absorbed.

The Mössbauer spectrum $I(\delta)$ is determined experimentally, but what the experimenter usually wants to know is the cross section $\sigma(E)$. If S(E) is already known, $\sigma(E)$ can be determined by one of two methods.

If C and B are also known already, then

$$\frac{I(\delta)-B}{C} \equiv f(\delta) = \int_{-\infty}^{\infty} S(\delta-E)\exp[-t\cdot\sigma(E)]dE \qquad (2)$$

and exp $[-t\cdot\sigma(E)]$ can be determined by deconvoluting the integral, as discused by Dibar-Ure and Flinn.[1]

The other method, which can be used even when S(E), B

205

and C are not known, is to carry out the integration using
sets of trial forms of $t \cdot \sigma(E)$, $S(E)$, and trial values of B
and C in a least-squares fitting process to find a set that
gives the best agreement with the experimentally determined
spectrum $I(\delta)$.

Elsewhere in this volume, Shenoy, Friedt, Maletta and
Ruby emphasize the importance of using realistic functions
for S and σ, and of carrying out the integration exactly.
We describe here a computer study of the speeds and ac-
curacies of a number of numerical methods for carrying out
both convolutions and deconvolutions.

DECONVOLUTION

Fourier transformation is the usual way of carrying
out a deconvolution.   The procedure is to use the property
that the transform of the convolution of two functions is
the same as the product of the transforms of the two functions.   Applied to Eq. (3), this gives

$$\mathcal{F}\{f(\delta)\} = \mathcal{F}\{S(\delta)\} \cdot \mathcal{F}\{\exp[-t \cdot \sigma(\delta)]\} \tag{3}$$

where the $\mathcal{F}$ is the Fourier transformation operator.   If we
divide both sides of Eq. (3) by the transform of the source
shape S and then antitransform, we obtain

$$\exp[-t \cdot \sigma(E)] = \mathcal{F}^{-1}\{\mathcal{F}[f(\delta)]/\mathcal{F}[S(\delta)]\} \tag{4}$$

where $\mathcal{F}^{-1}$ represents the inverse Fourier transformation.

By taking the logarithm of both sides of Eq. (4), the
form of $t \cdot \sigma(E)$ is found.   This may then be fitted by an
appropriate theoretical model with adjustable parameters.
The deconvolution is carried out just once.

1.   Fast Fourier Transform

Several techniques exist for carrying out Fourier
transformation numerically, but the most universally satisfactory one is the Fast Fourier Transformation which was
used by Dibar-Ure and Flinn.[1] To minimize the effects of
statistical noise in the experimental data they applied a
Gaussian filter to the transformed data, which is roughly

equivalent to smoothing the original data by taking a weighted average of each data point with its neighboring points.

2.   Fourier Transformation by the Hermite Expansion Method

We have also tested a newly developed numerical technique for carrying out Fourier transformations based on the properties of Hermite polynomials. The following discussion of the Hermite expansion method is based on the work of Afshar, Mueller and Shaffer.[2]

This procedure depends heavily on the fact that the Fourier transform of a Hermite function of order n is the same Hermite function multiplied by the factor i^n where $i = \sqrt{-1}$. Therefore, instead of spending time calculating the Fourier transform of a function it is possible to expand the function in Hermite polynomials and obtain the Fourier transform directly.

Suppose that two functions $G(\omega)$ and $H(\omega)$ are related through a Fourier transformation by

$$G(\omega) = \int_{-\infty}^{\infty} F(\omega') \exp(i\omega\omega') d\omega' \tag{5}$$

Simple numerical procedures exist for expanding G and F in any complete orthonormal set $Y_n(\omega)$, so that

$$G(\omega) = \sum_n a_n Y_n(\omega) \tag{6}$$

and

$$F(\omega) = \sum_n b_n Y_n(\omega) \tag{7}$$

When the $Y_n(\omega)$ are chosen to be the set of Hermite functions that are obtained from the Hermite polynomials $H_n(\omega)$ by

$$Y_n(\omega) = (2^n n! \sqrt{\pi})^{-\frac{1}{2}} H_n(\omega) \exp(-\omega^2/2) \tag{8}$$

it turns out that

$$b_n = i^n a_n \tag{9}$$

Fortunately, the calculational procedure is simplified by the ease with which the Hermite polynomials $H_n(\omega)$ can be calculated from the recursion formula

$$H_{n+1}(\omega) = 2\omega H_n(\omega) - 2nH_{n-1}(\omega) \tag{10}$$

Evaluation of the Fourier transform of G then requires the following two steps.

First, using the orthonormal property of the Hermite functions, it is easy to calculate

$$a_n = \int_{-\infty}^{\infty} G(\omega) Y_n(\omega) d\omega \quad , \tag{11}$$

where Simpson's rule can be applied for carrying out this integration reasonably quickly.

Next, the coefficients b_n for the Fourier transform are obtained from Eq. (9), and the Fourier transform is then evaluated.

3.   Exact Fourier Transform of the Source Line Shape

To save time and improve accuracy we have used the exact expression for the Fourier transform of a Lorentzian source line shape in all calculations involving the Hermite expansion method. That is,

$$\mathcal{F}[S(\delta)] = \frac{1}{\sqrt{2\pi}} \int_{-\infty}^{\infty} S(\delta) e^{i\alpha\delta} d\delta$$

$$= \frac{1}{\sqrt{2\pi}} \int_{-\infty}^{\infty} \frac{\Gamma_s/2\pi}{\delta^2 + [\Gamma_s/2]^2} e^{i\delta\alpha} d\delta$$

$$= \exp[-\alpha\Gamma_s/2] \quad . \tag{12}$$

Although Dibar-Ure and Flinn used a Lorentzian source line shape in their sample program, they used the Fast Fourier method to calculate its transform. Clearly the exact formula is an improvement over calculation by any approximate method. However, there are actual situations where the source line is not Lorentzian and this formula cannot be used.

4.   Comparison of the Two Deconvolution Methods

To test the accuracy of a numerical approximation procedure it is necessary to make a calculation for a case where the exact result is known. As a test we used the convolution of two Lorentzians

$$f(\delta) = \int_{-\infty}^{\infty} (\frac{\Gamma_s}{2\pi})/[(\delta-E)^2+(\Gamma_s/2)^2][1- \frac{t\sigma_0}{(E-E_a)^2+(\frac{\Gamma_a}{2})^2}]dE \quad (13)$$

because the exact formula for the integral is known to be

$$f(\delta) = 1- \frac{t\sigma_0(\Gamma_a+\Gamma_s)/\Gamma_a}{(\delta-E_a)^2+(\frac{\Gamma_a+\Gamma_s}{2})^2} \quad (14)$$

Of course the "transmission function"

$$1- t\cdot\sigma(E) = 1- \frac{t\sigma_0}{(E-E_a)^2+(\frac{\Gamma_a}{2})^2} \quad (15)$$

is not a realistic one, although it is often used in fitting Mössbauer data. The reason for using it here is that if the deconvolution procedure gives a good result for this test case it can be expected to give a good result for more realistic but more complicated transmission functions.

To make the test we calculated a 200-point mocked-up spectrum $f(\delta)$ from the exact formula of Eq. (14) and then used this calculated $f(\delta)$ and the source function

$$S(E) = \frac{\Gamma_s/2\pi}{(\delta-E)^2+(\frac{\Gamma_s}{2})^2} \quad (16)$$

as inputs to the deconvolution program. The output of the deconvolution program is supposed to be the transmission function, and it can be tested for accuracy by comparing it with the exact form given in Eq. (15).

Comparisons of the transmission functions calculated by the two different methods with the exact results are given in Fig. (1). The Fast Fourier Transform, at 3.6 seconds running time on the Northern Illinois University IBM 360-67, is obviously faster than the Hermite expansion method at 8.0 seconds. Since the deconvolution is carried out but once on a given set of data, time is not a critical element in this calculation.

The transmission line shape calculated using the Fast Fourier Transform agrees well with the known shape in the

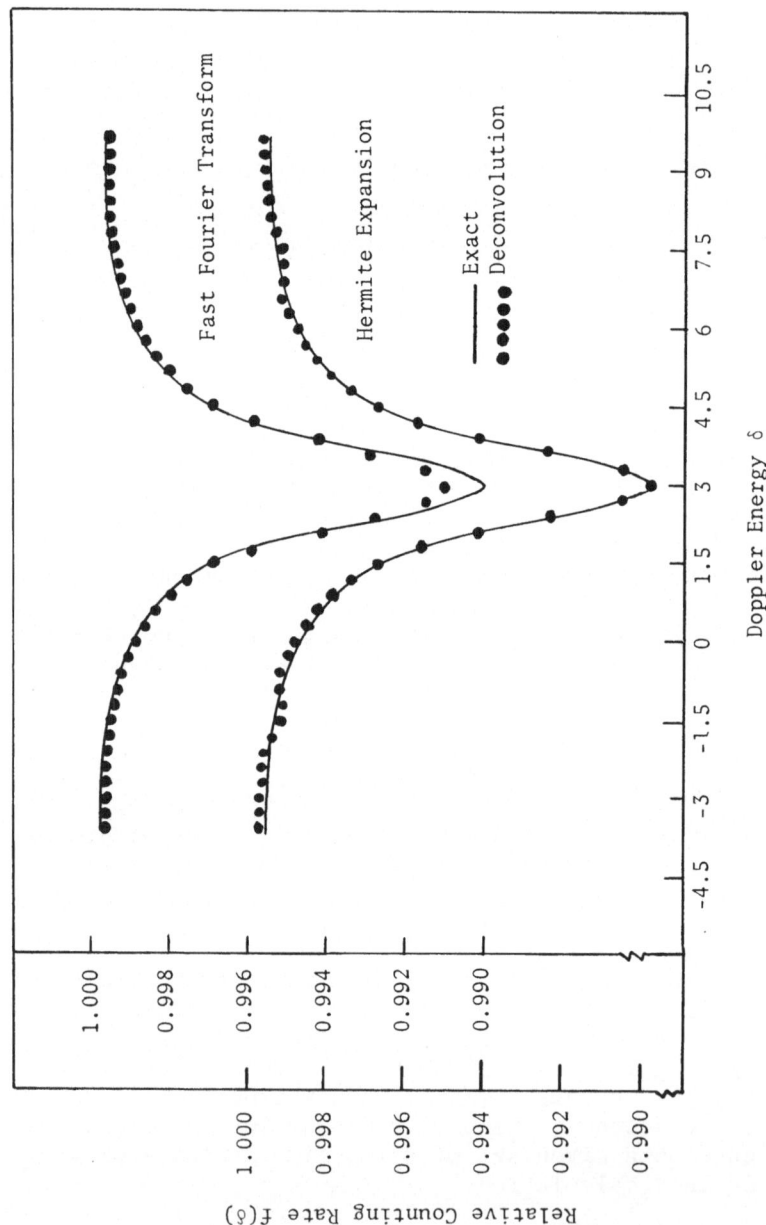

Figure 1   Deconvolution by Fourier Transform methods.

tails of the spectrum, but the intensity near the center does not. As noted by Dibar-Ure and Flinn, this fact is related to the use of the Gaussian filter.

On the other hand, the transmission function calculated by the Hermite expansion program fits almost perfectly at the center of the line, but poorly in the tails. In this calculation a high-frequency cut-off was used instead of a Gaussian filter. Perhaps some combination of the Fast Fourier Transform and the Hermite expansion methods would give more accurate overall results.

CONVOLUTION

Once a source function and a transmission function have been chosen there are many techniques for calculating the convolution integral. Six of the following seven methods were investigated. In each case a 200 point spectrum was calculated by evaluating the convolution integral 200 times using 200 different values of δ.

1. Exact Mathematical Expression for the Integral

The exponential form of the transmission function makes it impossible to find an exact expression for the convolution integral. But if the absorber is thin enough and the cross section is a sum of Lorentzians, the transmission function may be approximated by

$$T(E) = 1 - \sum \frac{t(\sigma_0)_i}{(E-E_i)^2 + (\frac{\Gamma a}{2})^2} \tag{17}$$

If the source function is also Lorentzian, the result of the convolution is as given in Eq. (14) if a summation sign is added before the last term on the right.

This approximation is widely used in fitting Mössbauer data. It requires only a fraction of a second to calculate a whole spectrum on our computer, but it leads to serious errors in least squares fitting of real data because the absorber is never ideally thin and the true source function may not be Lorentzian.

2. Approximate Expressions for Single Line Absorbers

A number of approximate procedures have been worked out for calculating the shapes of single resonances or well separated absorption lines when both the source and the absorber cross section are Lorentzian.[3,4,5] All of these methods introduce errors when used for thick absorbers with poorly resolved lines. We have not tested any of these methods since we are interested in procedures which can be used generally, without restrictions on the shapes of the source and transmission functions.

3. Numerical Integration

When there is no analytical expression for an integral that must be calculated, the most direct procedure is to perform a numerical integration.

Since any reasonable Mössbauer source function goes to zero at energies far from resonance, while the transmission goes to unity at energies far from resonance, it is usual to split the convolution integral

$$f(\delta) = \int_{-\infty}^{\infty} S(\delta-E)T(E)dE$$

into three parts,

$$f(\delta) = \int_{-\infty}^{-\ell} S(\delta-E)T(E)dE + \int_{-\ell}^{\ell} S(\delta-E)T(E)dT \int_{\ell}^{\infty} S(\delta-E)T(E)dE. \quad (18)$$

For $\ell \gg 1$, approximate formulas can be found for the first and third terms. Let us consider the first term for the case where the source shape is Lorentzian,

$$\int_{-\infty}^{-\ell} S(\delta-E)T(E)dE = \int_{-\infty}^{-\ell} \frac{\Gamma_s/2\pi}{(\delta-E)^2+(\Gamma_s/2)^2} dE. \quad (19)$$

For large enough values of ℓ, the transmission can be taken to be unity, independent of the details of the absorption cross section. Thus Eq. (19) can be simplified to

$$\int_{-\infty}^{-\ell} S(\delta-E)T(E)dE = \int_{-\infty}^{-\ell} \frac{\Gamma_s/2\pi}{(\delta-E)^2+(\Gamma_s/2)^2} dE$$

$$= (1/\pi) \{ \arctan [-\frac{2(\delta-\ell)}{\Gamma_s}] + \frac{\pi}{2} \} . \quad (20)$$

If the source function is not Lorentzian, a different approximation can be found. There will be a similar expression for the third term of Eq. (18). If we call the sum of the first and third terms $f'(\delta)$, then Eq. (18) becomes

$$f(\delta) = \int_{\ell}^{\ell} S(\delta-E)T(E)dE + f'(\delta) \ . \qquad (21)$$

To evaluate the first term of Eq. (21) the histogram approximation

$$\int_{-\ell}^{\ell} S(\delta-E)T(E)dE = \sum_{j=-\ell+\Delta E}^{\ell} S(\delta-E_j)T(E_j)\Delta E \qquad (22)$$

might be used. This term is the largest part of the total, and the accuracy of the calculation can be improved by using Simpson's rule or even something more elaborate instead of this rather crude approximation.

In a direct application of this method it took our computer 35 seconds to calculate a 200 point spectrum for a single absorption line. Since a least-squares fitting may require the calculation of hundreds of spectra, the computer time could make this an expensive calculation.

4. Cranshaw's Integration Procedure

This method, devised by T.E. Cranshaw,[6] is a clever application of the numerical integration method just described. The trick is to avoid the repeated recalculation of $S(\delta-E_j)$ and $T(E_j)$ for use in Eq. (22). This is accomplished by setting all the values of ΔE_j in Eq. (22) exactly equal to the fixed spacing between the successive velocity values δ for which $f(\delta)$ is to be calculated. That is, all the values of ΔE_j are set equal to $\Delta E = \delta_{k+1}-\delta_k$.

Thus Eq. (22) becomes

$$\int_{\ell}^{\ell} S(\delta_i-E)T(E)dE = \Delta E \sum_{j=-n/2}^{n/2} S(\delta_i-E_j)T(E_j) \ . \qquad (23)$$

But $E_j=j\Delta E$ and $\delta_i=i\Delta E$ so that

$$S(\delta_i-E_j) = S[(i-j)\Delta E] \qquad (24)$$

and

$$T(E_j) = T(j\Delta E) \qquad (24')$$

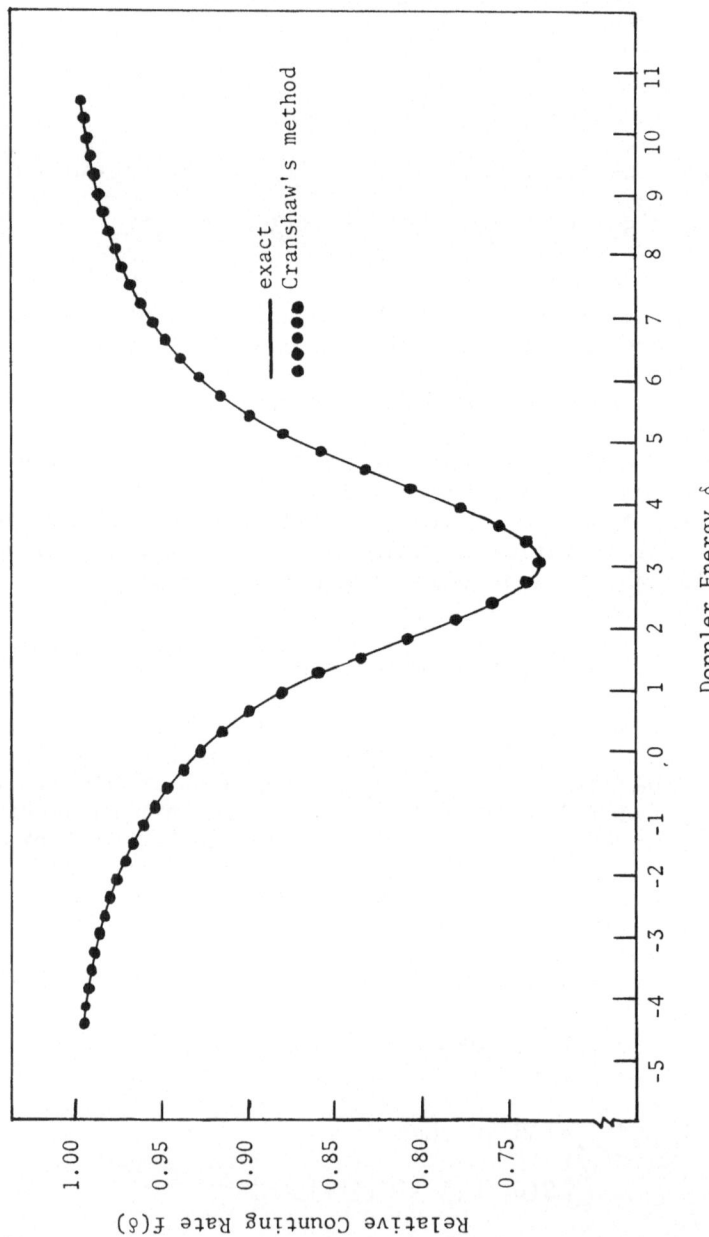

Figure 2   Convolution integration by Cranshaw's method.

and Eq. (22) can be written as

$$\int_{-\ell}^{\ell} S(\delta_i - E)T(E)dE = \Delta E \sum_{j=-n/2}^{n/2} S[(i-j)\Delta E]T(j\Delta E) \qquad (25)$$

At the very beginning of the computer program, two one-dimensional arrays (vectors) are calculated. The elements of these vectors are

$$S_j = S(j\Delta E)$$

for the source, and

$$T_j = T(j\Delta E)$$

for the transmission. For each new value of δ_i the elements T_j of the transmission vector remain unchanged, while the corresponding elements S_{i-j} of the source vector for Eq. (25) are obtained immediately from the values of S_k for $k=i-j$ with no new calculation of S. Thus for each point in the calculated spectrum only the products and sum in Eq. (25) must be calculated. The time-consuming repeated calculations of the functions S and T are avoided.

Cranshaw[6] has used this integration procedure to take into account not only the finite absorber thickness, but also a number of other troublesome experimental effects such as inhonogeneous broadening of the source spectrum and the consequences of poor geometry.

For a 200 point spectrum with n=600 in Eq. (25), the total calculating time was 2.9 sec. None of Cranshaw's other corrections were made. Figure (2) compares the result of Cranshaw's method with the exact result when both are calculated for the thin absorber approximation.

5. Gaussian Integration

A special modification of the Gaussian method of integration[7] constructed by B. Dunlap[8] turns out to be the fastest numerical integration procedure if there is only one absorption line. When applied to a multiple-line spectrum the computing time increases. As now used, this procedure requires that the source function be Lorentzian.

Following a procedure attributed to Czjzek,[9] Dunlap

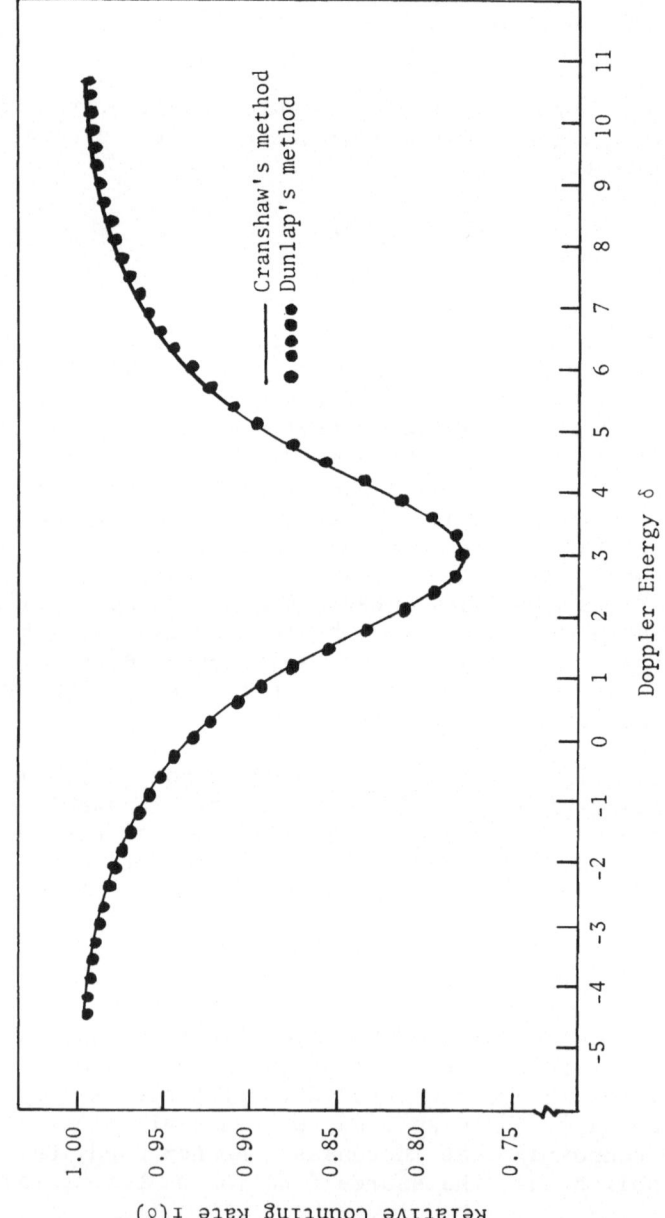

Figure 3   Convolution integrations.   Dunlap's method compared to modification of
Cranshaw's method.

separates the integral

$$f(\delta) = \int_{-\infty}^{\infty} \frac{\Gamma_s/2\pi}{(\delta-E)^2+(\Gamma_s/2)^2} \exp\left[-\frac{t\sigma_0}{(E-E_a)^2+(\Gamma_a/2)^2}\right] dE$$

$$= \int_{-\infty}^{\infty} S(\delta-E) \exp\left[-\frac{t\sigma_0}{(E-E_a)^2+(\Gamma_a/2)^2}\right] dE \qquad (26)$$

into two parts,

$$f(\delta) = f_1(\delta)+f_2(\delta) \quad , \qquad (27)$$

where

$$f_1(\delta) = \int_{-\infty}^{\infty} S(\delta-E) \left[1-\frac{t\sigma_0}{(E-E_a)^2+(\Gamma_a/2)^2}\right] dE \qquad (28)$$

and

$$f_2(\delta) = \int_{-\infty}^{\infty} S(\delta-E) \left\{\exp\left[-\frac{t\sigma_0}{(E-E_a)^2+(\Gamma_a/2)^2}\right]-\left[1-\frac{t\sigma_0}{(E-E_a)^2+(\Gamma_a/2)^2}\right]\right\} dE.$$
$$(29)$$

The first part, f_1, is a close approximation to the original integral, and it can be evaluated directly by use of the known analytical expression for this integral, Eq. (14).

The second part, f_2, is a relatively small correction term whose integrand approaches zero rapidly and smoothly for large positive and negative values of E. Thus an approximation method can be used to evaluate this correction with good accuracy. It is the integral f_2 which Dunlap evaluates by Gaussian integration. In this well known numerical integration procedure accuracy is improved by giving up the usual requirement of equally spaced abscissas and a fixed value of ΔE. Instead, the integrand is evaluated at values of E chosen in such a way as to maximize the accuracy of the result.

The total time required to calculate a 200 point spectrum with one absorption line was 1.9 seconds. For additional absorption lines, additional time was required. Thus a calculation for three lines took 2.4 seconds. Figure (3) shows a comparison of the result of this method with the result obtained by Cranshaw's method for a thick absorber.

6. Modification of Cranshaw's Method

If the source line shape is Lorentzian, the separation of the integral into a part that can be evaluated exactly and a correction that is evaluated by numerical integration can be applied to Cranshaw's method. The computing time is about the same as for the unmodified procedure. When this modification is used, the "wing correction" $f'(\delta)$ in Eq. (21) must be omitted.

7. Integration by the Fourier Transform Method

The Fourier transform methods used for deconvolution can be used for convolution. The integral

$$f(\delta) = \int_{-\infty}^{\infty} S(\delta-E)T(E)\,dE \tag{30}$$

is given by

$$f(\delta) = \mathcal{F}^{-1}\{\mathcal{F}[S(E)] \cdot \mathcal{F}[T(E)]\} \tag{31}$$

Some of the problems encountered in the deconvolution process are not so serious here. Chief among these is the one caused by the division of one frequency spectrum by another in the deconvolution process. This produces violent fluctuations in the quotient at high frequencies because small numbers with large relative uncertainties are divided by other small numbers with large relative uncertainties. These uncertainties are partly a consequence of the numerical procedure, although they are worse if a real Mössbauer pattern with statistical fluctuations in the count rate is used in the calculation. The reason for using the Gaussian filter in the deconvolution calculation is to suppress these fluctuations.

In carrying out a convolution, the trouble-making division process is replaced by a multiplication, and both of the functions in the integrand are smooth functions. Thus the fluctuation problem is much less serious. In addition, if the source function is Lorentzian, its exact transform is given by Eq. (12), and need not be calculated numerically. Thus one source of error due to the inexact numerical procedure can be eliminated.

When the Fast Fourier Transform was used, the computing

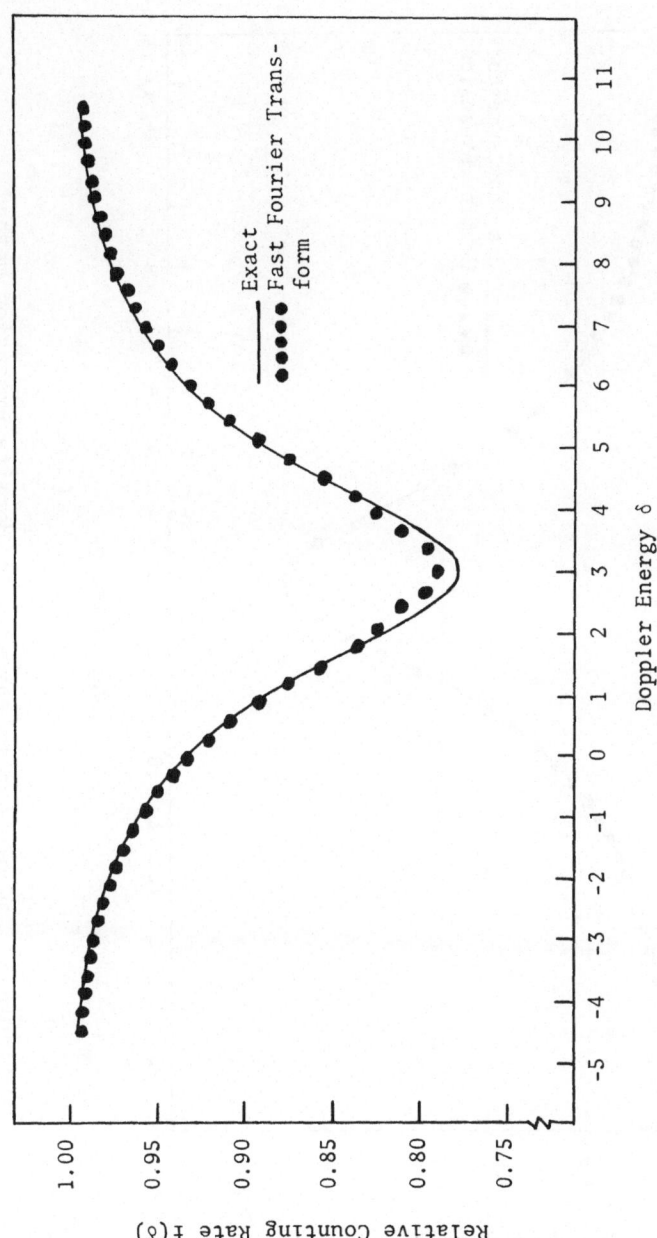

Figure 4   Convolution integration by Fast Fourier Transform method.

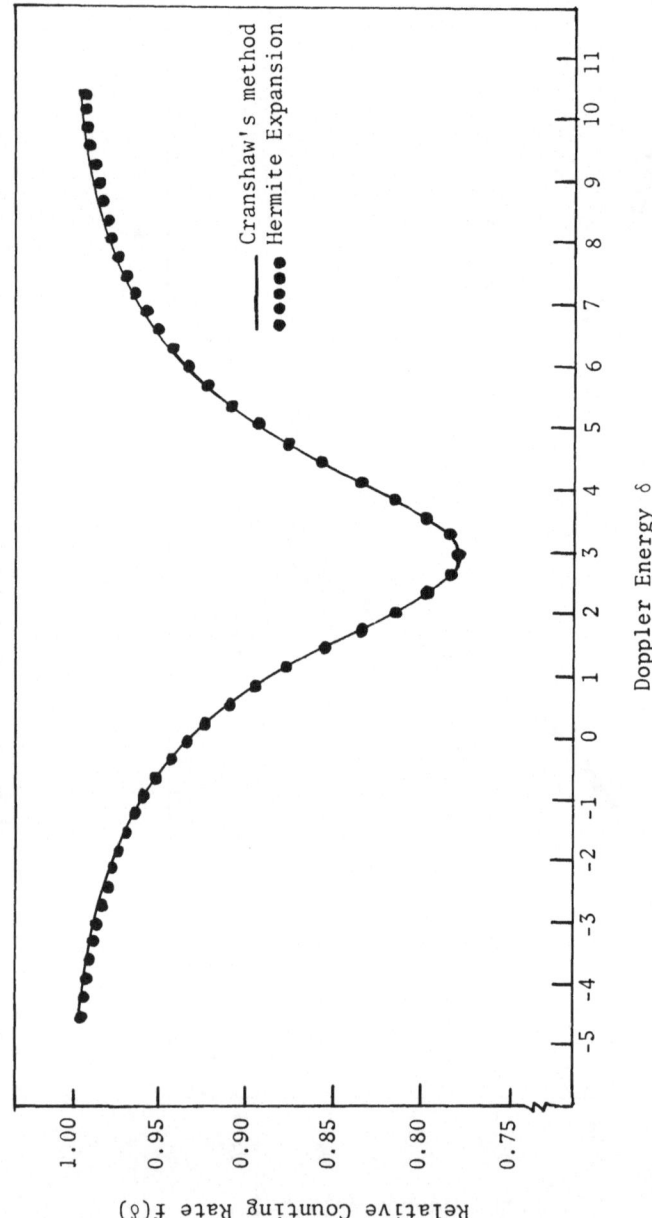

Figure 5   Convolution integration by Hermite Expansion method.

time was 3.12 seconds. Figure (4) shows a comparison of
the result of this method with the exact result when both
are calculated for the thin absorber approximation.

When the Hermite expansion method was used the computing time was 7.4 sec. A comparison of the result of
this calculation with the result obtained by Cranshaw's
method is shown in Figure (5).

7. Comparisons of Accuracy

If a least-squares fit is used to get the shape of the
absorption cross section, the convolution integral will
have to be evaluated many times for different sets of values
of the adjustable parameters. Therefore the speed of calculation of the integral is very important. But the convolution procedure must be sufficiently accurate, as well
as fast, or it cannot be used at all.

Except for Dunlap's method and the modification of
Cranshaw's method, all of the convolution procedures could
be tested for accuracy by using the thin absorber approximation given in Eq. (13), and comparing the computed result
with the exact result, Eq. (14). The accuracy of the other
two methods had to be evaluated indirectly.

The maximum error in the result by Cranshaw's method
was approximately 10^{-5} which is of no importance for any
practical purpose.

Having acquired confidence in the accuracy of Cranshaw's method, we applied it to the calculation of a convolution integral for the case where the transmission was
an exponential function instead of the thin absorber
approximation. We then calculated a spectrum for the same
transmission function using Dunlap's method. On the
assumption that the result of Cranshaw's method has an error
of $\sim 10^{-5}$, the error in Dunlap's result is about 10^{-4} at
maximum.

The same spectrum was also calculated by the modified
version of Cranshaw's method with results that differed by
$\sim 10^{-5}$ from the result of the unmodified version. This
increased our confidence in the accuracy of Cranshaw's
method.

TABLE I

COMPARISON OF METHODS FOR EVALUATING
THE CONVOLUTION INTEGRAL

Method	Time (Seconds)	Worst Error (Fractional)
Thin absorber, exact expression	Very fast	------
Standard numerical integration	35	$3 \cdot 10^{-5}$
Cranshaw's method		
1 line	2.8	$3 \cdot 10^{-5}$
3 lines	3.0	$3 \cdot 10^{-5}$
Dunlap's method		
1 line	1.9	10^{-3}
3 lines	2.4	10^{-3}
Fast Fourier Transform	3.1	10^{-2}
Hermite expansion	7.4	10^{-4}

A comparison of the different convolution methods is given in Table I. Clearly, Cranshaw's method gives the greatest accuracy at almost the highest speed. A sample Fortran program for this method is given in the appendix. With some computers it may be possible to shorten the calculation time by using a lower level programming language in evaluating Equation (25) if this will eliminate the arithmetic operation of actually calculating i-j (the Fortran instruction labeled 401 in our program).[10]

CONCLUSIONS

For deconvolution the Fast Fourier Transform is faster than the Hermite expansion method and gives a smoother curve.

For convolution Cranshaw's method is the most accurate and very fast.

REFERENCES

1. M. Celia Dibar-Ure and P.A. Flinn, "Mossbauer Effect Methodology", Vol. 7, I.J. Gruverman, Ed., p. 245. Plenum Press, New York (1971). References 7 - 10 in this paper are descriptions of the Fast Fourier Transform technique.

2. R. Afshar, F. Mueller and J.C. Shaffer, J. Comp. Phys. 11, 190 (1973).

3. S. Margulies and J.R. Ehrman, Nuclear Inst. and Meth. 12, 131 (1961).

4. S.L. Ruby and J.M. Hicks, Rev. Sci. Instr. 33, 27 (1962).

5. J. Heberle and S. Franco, Z. Naturforsch. 23, 1439 (1968).

6. T.E. Cranshaw, Journal of Physics E 7, 1 (1974).

7. "Handbook of Mathematical Functions", M. Abramowitz and I.A. Segun, Eds., p. 887. Dover Publications, Inc., New York, 1965.

8. B. Dunlap, private communication.

9. G. Czjzek, private communication to B. Dunlap.

10. T.E. Cranshaw, private communication.

APPENDIX

```
C CONVOLUTION INTEGRATION BY CRANSHAW'S METHOD
      IMPLICIT REAL*8 (A-H,O-Z)
      DIMENSION F(301),T(1001),V(10),A(10),S(1001)
      VELR=30.  Establish velocity range of spectrum.
      XL=30.  Extension of range for numerical integration.
      API=3.1415926
      GAMS=2.   Source line width.
      GAMA=2.   Absorber line width.
      NJ=3      Number of absorption lines.
      VEL=VELR/2.
      X=-VEL
      NP=200    Number of velocity intervals in spectrum.
      DELTE=VELR/NP  Width of velocity interval.
      A(1)=.5
      A(2)=.3   fnσ for absorption lines.
      A(3)=.1
      V(1)=3.
      V(2)=0    Positions of absorption lines.
      V(3)=-4.
      NP1=NP+1  Number of points in spectrum.
      L=(XL+VEL)*2.*NP/VELR+1.1
      EN=-VEL-XL-DELTE
      DO 100 I=1,L   Calculate the source function S.
      EN=EN+DELTE
  100 S(I)=GAMS/2/API/(EN**2+GAMS**2/4.)
      X=-XL-DELTE
      L=L-200
      DO 102 I=1,L
      T(I)=0.   Calculate the transmission function T.
      X=X+DELTE
      DO 101 J=1,NJ
  101 T(I)=T(I)+A(J)/((X-V(J))**2+(GAMA/2.)**2)
  102 T(I)=DEXP(-T(I))
      EN=-VEL-DELTE
      LL=L-1
      DO 400 I=1,NP1
      EN=EN+DELTE              "Wing" correction.
      COR=1./API*(API-DATAN(2.*(XL+EN)/GAMS)
     1-DATAN(2.*(XL-EN)/GAMS))
      F(I)=(S(202-I)*T(1)+S(201+L-I)*T(L))/2.
      DO 401 J=2,LL           Calculate the sum in Eq. 25.
  401 F(I)=F(I)+S(201+J-I)*T(J)
      F(I)=F(I)*DELTE+COR   Multiply the sum by ΔE and
  400 PRINT 3,EN,F(I)       add the wing correction.
    3 FORMAT(4(1X,F14.8))
      END
```

DETERMINATION OF TRANSMISSION INTEGRAL PARAMETERS BY
SIMULTANEOUS FITS TO THICK-ABSORBER MÖSSBAUER SPECTRA:
APPLICATION TO $^{151}Eu_2O_3$

David G. Agresti[†], Michael Belton[†],
John Webb[††], and Saphura Long[†]

[†]Department of Physics and
[††]Division of Immunology and
Rheumatology, Department of Medicine

University of Alabama in Birmingham
Birmingham, Alabama 35294

ABSTRACT

Thick-absorber line shape analysis has been combined
with a least-squares procedure capable of simultaneously
fitting complementary Mössbauer spectra. The technique has
proved sufficiently powerful that convergence has been
achieved for the first time in a single least-squares fit
with every parameter of the transmission integral formulation free to vary. In applying the method, transmission data
were collected for nine Eu_2O_3 absorbers distributed in thickness from 4 to 100 mg Eu/cm^2. The spectra were then fit
simultaneously in two independent groups, and the parameter
values obtained from the two fits were consistent to within
statistical error. Results for the recoil-free fractions
are: $f_s(^{151}EuF_3) = 0.485 \pm 0.005$ and $f_a(Eu_2O_3) = 0.546 \pm
0.016$ (if $\alpha = 28.5$); and for the effective quadrupole split-
ting parameter: $e^2qQ = -5.11 \pm 0.36$ mm/sec.

I.   INTRODUCTION

In Mössbauer spectra that exhibit unresolved magnetic or
quadrupole interaction, fitting the spectra with the Lorentzian approximation yields limited information about the
chemical environment of the resonant nucleus. Although the
isomer shift is usually fitted to good precision, this is
not normally true of the parameters that characterize the
transmission integral, such as the recoil-free fraction of
the source or the absorber. Moreover, even the fitted value
of the isomer shift may be suspect for ^{151}Eu spectra if unresolved quadrupole broadening is present [1].

Recently, we have begun a study of europium as a
calcium substituent in biological complexes. Because of the
chemical similarity of the calcium and europium ions,
Mössbauer spectroscopy in ^{151}Eu is expected to be a suitable spectroscopic tool for such studies [2]. However,
room temperature Mössbauer spectra are generally unresolved.
Consequently, in order to expand the information that may
be derived from such spectra, we have used an analysis that
combines a calculation of the true spectrum shape (transmission integral formulation) with the advantages of simultaneous fitting of complementary spectra.

The desirability of simultaneous   fitting of complementary spectra is not generally realized. Furthermore,
we will show with this paper that simultaneous fitting of
data is essential if transmission spectra are to be fully
characterized. In this work we have applied the technique
to unresolved ^{151}Eu$_2$O$_3$ transmission spectra and have suc-
ceeded in determining to high precision the parameters of
the transmission integral formulation as well as the quadrupole splitting parameter.

II.   EXPERIMENTAL DETAILS

A 30 mCi source of ^{151}Sm embedded in anhydrous EuF$_3$
(New England Nuclear) was chosen for its relatively narrow
emission spectrum [3]. The 21.54 keV [4,5], 7/2+ → 5/2+,
first excited to ground-state transition in ^{151}Eu was
detected with a 30 mm^2 x 3 mm cooled Si(Li) detector.
Acceptable counting rates, which varied with absorber thickness from 60,000 to 2,000/day over 100 channels, were obtained with a constant source-to-detector distance of approxi-

Table 1. The nine Eu_2O_3 absorbers of thickness t, and the results for N (∞), ε, and Γ_{exp} of single-line Lorentzian fits to the nine transmission spectra of Fig. 1. Resonance effect, ε, is the fitted relative absorption of the spectrum divided by the background correction, q_∞.

#	t ($mgEu/cm^2$)	q_∞	N(∞)	ε (%)	Γ_{exp} (mm/sec)
1	3.87	0.928	121.7K	12.89±0.10	2.84±0.05
2	8.20	0.941	48.9K	20.47±0.14	3.35±0.05
3	15.53	0.936	37.8K	29.85±0.15	3.40±0.04
4	26.8	0.927	82.6K	36.32±0.10	3.78±0.03
5	36.1	0.939	55.4K	38.73±0.11	4.31±0.03
6	52.2	0.930	60.6K	40.49±0.12	4.51±0.03
7	68.7	0.920	9.4K	42.14±0.30	5.17±0.09
8	78.3	0.933	5.1K	42.09±0.37	5.57±0.13
9	100.6	0.928	9.6K	42.30±0.31	5.59±0.09

mately 1.3 cm. The small β-branching ratio of 2% to the 21.54 keV level and the large internal conversion coefficient of 28.5 [5,6] reduced the effective γ-ray activity to about 20 μCi and necessitated the use of a comparatively large solid angle. The velocity transducer was of the Kankeleit type [7] and calibrated to better than 0.5% with reference to natural iron or iron oxide, as appropriate.

Nine absorbers (listed in Table 1) of various thicknesses in Eu_2O_3 (Alfa Inorganics, 99.99%) were prepared by dispersing a weighed amount of fine oxide powder (particle dia. $\simeq$ 1.5-2.5μ) in melted paraffin and shaping with a 5/16" dia. pill press. Errors in the absorber thickness are estima-

ted to range from ±2% for the thinnest absorber to ±0.5% for
the thickest. Granularity [8] may affect the spectra for
the two thinner absorbers, but should be unimportant for absorbers with thickness > 15 mg Eu/cm^2 (around 12 particles
thick). Debye-Scherrer powder patterns, obtained using MoK$_\alpha$
x radiation, were in agreement in line positions and relative intensities from absorber to absorber and with the
ASTM Powder Diffraction File entry for cubic Eu$_2$O$_3$(a =
10.869 Å) [9].

III. METHOD OF ANALYSIS

It is well known that the proper theoretical function
for thick-absorber transmission spectra is that given by
Marqulies and Ehrman in 1961 [10]. Thus, the number of
counts, N, recorded as a function of the source velocity, v_s,
may be written in a form useful for the present work as

$$N(v_s) = N(\infty) \cdot [(1-q_\infty f_s) + q_\infty f_s \cdot I(v_s)], \qquad (1a)$$

where the transmission integral, $I(v_s)$, is given by

$$I(v_s) = \frac{2}{\pi \Gamma_s} \int_{-\infty}^{\infty} dv \, \frac{\exp\left[-\left(\frac{t}{t_M}\right) \sum \frac{h_j}{4(v-v_j)^2 / \Gamma_\alpha + 1}\right]}{4(v-v_s)^2 / \Gamma_s^2 + 1}, \qquad (1b)$$

and $N(\infty)$ represents the number of counts in the background.
In this form, we assume a Lorentzian emission spectrum of
width, Γ_s; that is, we assume the thin source approximation
with possible unresolved Lorentzian broadening (if $\Gamma_s > \hbar/\tau =$
0.65 ± 0.02 mm/sec [11]).

The absorption spectrum is taken to be a sum of Lorentzian components, each with width, Γ_a, velocity position v_j,
and relative intensity h_j, where $\Sigma h_j = 1$. Since there is
evidence [3,12] for quadrupole splitting in the absorber,
the v_j are given in terms of the axial component of the
electric field gradient, eq, the excited and ground-state
nuclear quadrupole moments, eQ_{ex} and eQ_{gd}, the asymmetry
parameter, η, and the isomer shift, δ. The ratio of quadrupole moments, Q_{ex}/Q_{gd}, has been determined to good precision
by several groups [3,13], and we have adopted the value,
$Q_{ex}/Q_{gd} = 1.30$ [11]. If, as is frequently done for poorly
resolved spectra, the assumption is made that $\eta = 0$, then the
v_j depend on just two adjustable paramters, δ and e^2qQ, the

"effective" quadrupole splitting parameter for the ground
state. There would then be 8 lines (for a pure M1 transition [5]) in the absorption spectrum with line intensities,
h_j, proportional to the squares of the appropriate Clebsch-
Gordan coefficients.

The transmission spectrum depends as well on the recoil-
free fractions of the source, f_s, and the absorber, f_a, thru
the parameter, t_M, the Mössbauer absorption length, which
may be defined by [14]

$$t_M = \frac{A(1 + \alpha)}{N_o a \sigma_o f_a} \qquad (2)$$

where A = 151.96, α = 28.5, N_o = Avogadro's number,
a = 0.4782, and σ_o = 7.031 x 10^{-18} cm^2[15]. The remaining
two quantities in Eq. (1), t and q_∞, reflect experimental
conditions, since t represents the thickness of the absorber,
say in mils or in mg/cm^2, and q_∞ represents the "quality"
of the pulse height spectrum of the resonant γ-ray, namely
the peak-to-total ratio for v → ∞, and serves as a background
correction factor (See Table 1).

In principle, to fit a spectrum with the transmission
integral formation, q_∞, t, Q_{ex}/Q_{gd} and the set of values for
v_S are read in as preset quantities, while the parameters,
$N(\infty)$, f_{s}, t_M, Γ_s, Γ_a, δ, and e^2qQ, are varied until a minimum in χ^2 is reached. In practice, it has not proved possi-
ble, and predictions based on simulated data [16,17] show
that it is not likely that all these parameters can be
determined by fitting a single transmission spectrum. The
reason for this can be seen by comparing a Lorentzian spectrum,

$$N(v_S), \text{ Lor. } = N(\infty) \cdot \left[1 - q_\infty \varepsilon \sum \frac{h_j}{4(v-v_j)^2/\Gamma_{exp}^2 + 1} \right], \quad (3)$$

with the thick-absorber spectrum, Eq. (1). Four parameters
of the transmission integral formulation, f_s, t_M, Γ_s, and
Γ_a, have been replaced by just two parameters in the thin-
absorber approximation, namely resonance effect, ε, and
full width at half maximum, Γ_{exp}. It is always fairly easy
(especially if e^2qQ = 0) to obtain a good fit to a poorly
resolved spectrum with Eq. (3). In fitting with Eq. (1),
f_s is usually well determined, since $N(v_S)$ is linear in f_s,

but generally only one of the three parameters, t_M, Γ_s, and Γ_a, may be determined with good precision [17]. Evidently, the correlations among t_M, Γ_s, and Γ_a for a single transmission spectrum are too great to allow an unambiguous determination of their values. Therefore, to fit spectra with Eq. (1), either a way must be found to reduce these correlations, or one or more of the parameters must be given fixed values or related to another parameter.

In a 1971 article, Erickson et al. [18] reported a successful procedure for fitting ^{197}Au spectra with the transmission integral formulation. Their method was to obtain two spectra, one for a thin absorber and one for a thick absorber. They first deleted one variable parameter by setting $\Gamma_a = \Gamma_s$, an assumption well-justified for the naturally broad line in ^{197}Au. Then they fit the spectra individually with t_M variable and Γ_s fixed for the thick-absorber spectum and the reverse for the thin-absorber spectrum. They refit each spectrum, iterating until consistent values for t_M and Γ_s were obtained. In this way, they effectively reduced the correlations between these two parameters and obtained precise values (<1% error) for both t_M and Γ_s.

The present work uses basically the same procedure as that described by Erickson et al., except that we have employed the method of simultaneously fitting complementary spectra [19,20], rather than the iterative procedure used in the ^{197}Au work. That is, several spectra are fit together, and the parameters are varied to minimize the "sum" of the individual χ^2. More precisely, we define [20]

$$\chi^2_{Sim} = \frac{1}{N_s \cdot N_d - N_p} \sum_{s=i}^{N_s} \sum_{i=1}^{N_d} \frac{\left[Fs(xs_i) - Ys_i \right]^2}{Ys_i} \tag{4}$$

where N_s experimental spectra, Ys_i, are fit together with N_d channels of data per spectrum. The mean velocity for each channel, xs_i, may vary from spectrum to spectrum, e.g. as the maximum velocity setting is increased to accomodate the broadened spectra obtained with thicker absorbers. The fitted function, $Fs(xs_i)$, also varies from spectrum to spectrum, e.g. as t or $\bar{N}(\infty)$ is changed. In the present work, Fs depends as well on the parameters f_s, t_M, Γ_s, Γ_a, δ, and e^2qQ, which are constrained in the simultaneous fits to be the same for all spectra. As a result, there are N_p independently variable parameters in the simultaneous fit.

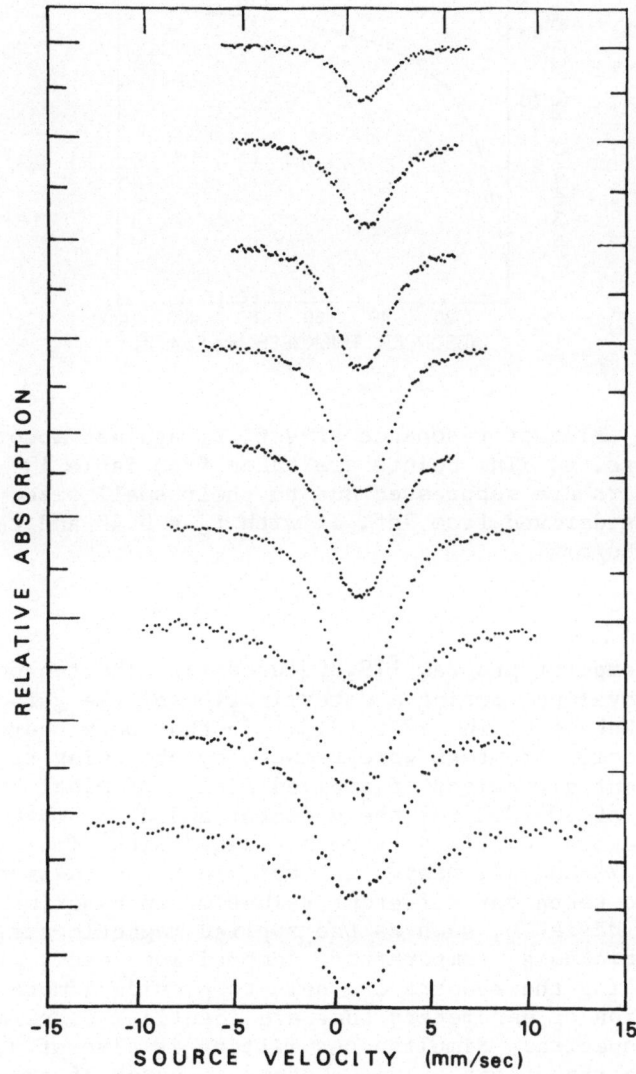

Fig. 1. Transmission spectra for the nine Eu_2O_3 absorbers. The spectra are arrayed from top to bottom in order of increasing absorber thickness (See Table 1). The scale on the vertical axis is 10% per minor division.

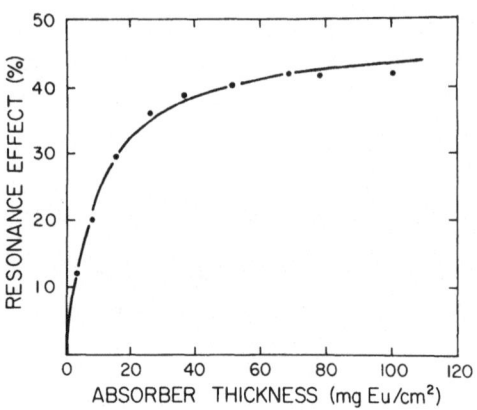

Fig. 2. Plot of resonance effect, ε, against absorber thickness, t. The points are taken from Table 1; error bars are suppressed due to their small size. The curve is derived from Ref. 16 with f_s = 0.48 and t_M = 5.5 mg Eu/cm^2.

    The computer program [19,20] used for simultaneous fitting was developed during a determination of the g-factor of the 99-keV level in ^{195}Pt [21]. In that work, correlations among variable paramters were reduced by obtaining spectra for different directions of applied field. Nominal standard deviations of 50-100% for the g-factor and H_{int}, obtained by fitting spectra individually were reduced with simultaneous fitting to 7% and 3%, resp. In fact, in any situation where spectra are taken for different values of an experimentally controlled variable, such as the applied magnetic field, or absorber thickness, temperature, composition, etc., simultaneously fitting the spectra can help to provide a more precise determination of parameters that are identical or related in different spectra. Simultaneous fitting is also of value in reducing the effect of any systematic error of experimental origin, such as that due to velocity nonlinearity, granularity or a large solid angle, that in the simultaneous fit itself is not strongly correlated to one of the variable parameters.

    For the numerical evaluation of the transmission integral, the relevant portions of the Fortran IV program [22] used by Erickson et al. [18] have been incorporated

as subroutines [23] into the simultaneous fitting program
[19,20]. The method employs a 192 point Gauss-Legendre quadrature [24] and requires less than 2 sec per integration on the
IBM 370/158. Execution time for 100-channel spectra was
about 16 sec per spectrum per iteration, and typically 5
iterations were allowed for each simultaneous fit.

IV. RESULTS

Transmission spectra recorded at room temperature (23–
25°C) for each of the nine Eu_2O_3 absorbers are shown in Fig.
1. Each spectrum consists of 100 data points obtained by
summing the counts in pairs of complementary channels.

As a guide to selecting the proper starting values for
the transmission integral parameters, the spectra were first
fit individually with the Lorentzian approximation, Eq. (3),
since the parameters ε and Γ_{exp} more nearly characterize the
spectrum as observed than the parameters f_s, t_M, Γ_s, and Γ_a.

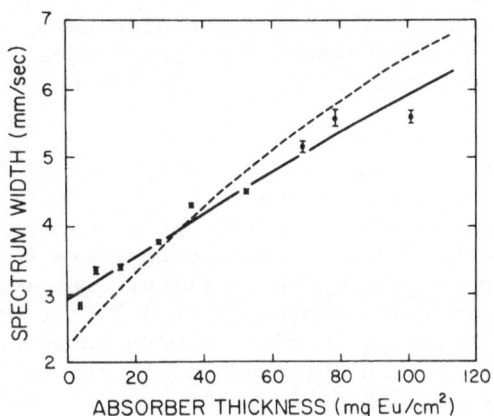

Fig. 3. Plot of spectrum width, namely full width at
half maximum, Γ_{exp}, against absorber thickness, t. The
points, with error bars, are taken from Table 1. The
solid curve is derived from Ref. 16 with $\Gamma_a = \Gamma_s = 1.45$
mm/sec and $t_M = 11.2$ mg Eu/cm^2. The dashed curve is
derived from Ref. 16 with $\Gamma_s = 1.12$ mm/sec and
$t_M = 5.48$ mg Eu/cm^2.

The results of the Lorentzian fits are presented in Table 1. In Figs. 2 and 3 we have plotted the observed variation of ϵ and Γ_{exp} with absorber thickness. The simplest approach to obtaining a set of starting values is to read off $f_s = 0.42$ a: the upper limit of the ϵ data, $\Gamma_s = 1.45$ mm/sec from the Γ_{exp} data extrapolated to zero thickness, and $t_M = 11.2$ mg Eu/cm^2 from the 13.5% rule [25] for increase of Γ_{exp} per unit increase in effective thickness, $T_a = t/t_M$, which assumes $\Gamma_a = \Gamma_s$. An alternative, and possibly more reliable procedure, is to assume $\Gamma_a = \Gamma_s$ for simplicity and fit theoretical curves [16] for the variation with T_a of ϵ/f_s and Γ_{exp}/Γ_s to the observed variation. Such a fit to the ϵ data gives $f_s = 0.48$ and $t_M = 5.5$ mg Eu/cm^2. The fit to the Γ_{exp} data gives the values for Γ_s and t_M obtained earlier. The fitted curves are displayed in Figs. 2 and 3 as the solid lines. Thus, as starting values to be used in the simultaneous fits, we may take $f_s = 0.48$, $\Gamma_a = \Gamma_s = 1.45$ mm/sec, and $t_M = 5.5$ or 11.2 mg Eu/cm^2.

Before performing the simultaneous fits, the spectra were grouped into two non-intersecting sets extending over the full range of absorber thickness. The purpose of this was to try to rule out artificial minima in χ^2 by requiring that the parameter values obtained from the two sets of data be consistent. Therefore the nine Eu$_2$O$_3$ spectra were grouped into an odd-numbered set, consisting of 5 absorbers, and an even-numbered set, consisting of 4 absorbers (See Table 1).

The first simultaneous fits were performed by setting the parameter $e^2qQ = 0$, since the Eu$_2$O$_3$ spectra were completely unresolved (See Fig. 1), and we were eager to assure convergence by not adding parameters for which the spectra seemed to have no information. The results of these initial fits are shown in Table 2. If we compare the parameter values of Simfit Ia with those of Simfit Ib, we see that the values obtained for f_s are in reasonable agreement, but the values for t_M, Γ_s, and Γ_a are inconsistent. To reduce still further the number of variable parameters, we repeated the fits with the constrain that $\Gamma_a = \Gamma_s$, since this possibility was not ruled out by Simfit I. We obtained the results shown as Simfits IIa and IIb, with marked improvement in the agreement among the parameters. Thus, reducing the number of variable parameters improved the consistency between the odd and even-numbered spectra. Nevertheless, the χ^2 values were rather large (2.59 and 3.83 and for Simfits IIa and IIb, resp.), and when we plot the expected variation with t of ε and Γ_{exp} [16] on the same graph as the observed variation (Figs. 2 and 3), the observed variation in ε is well predicted, but the slope of the predicted variation in Γ_{exp} is twice the observed slope.

In an attempt to reconcile these inconsistencies, we added e^2qQ as a variable parameter, using $e^2qQ = -4mm/sec$ [12] as the starting value. In view of our previous experience with Simfits I and II, it seemed very unlikely that we would obtain consistent values if we freed all the parameters, and our first simultaneous fits were done with f_s and t_M fixed to the average of the values of Simfit II. We also replaced Γ_a by $\Gamma_a + \Gamma_s$ as a variable parameter, since $\Gamma_a + \Gamma_s$ turned out to be less correlated to other

Table 2. Results of simultaneous fits with the transmission integral formulation to the nine transmission spectra (Fig. 1) with the constraint that $e^2qQ = 0$. In Simfits IIa and IIb, the additional constraint was imposed that $\Gamma_a = \Gamma_s$. The value $\delta = 1.000 \pm 0.003$ was obtained as the average of Simfits IIa and IIb.

Simfits	Absorbers[a]	f_s	t_M[b]	Γ_s[c]	Γ_a[c]
Ia	1,3,5,7,9	0.498	5.23	1.26	1.07
		±.006	±.16	±.05	±.03
Ib	2,4,6,8	0.486	6.32	1.10	1.20
		±.007	±.25	±.06	±.04
IIa	1,3,5,7,9	0.483	5.46	1.14	$= \Gamma_s$
		±.004	±.06	±.01	
IIb	2,4,6,8	0.475	5.51	1.09	$= \Gamma_s$
		±.005	±.08	±.02	
	Average	0.479	5.48	1.12	
		±.003	±.05	±.01	

[a] See Table 1

[b] mg Eu/cm^2

[c] mm/sec

Table 3. Results of the final transmission integral fits to the nine transmission spectra (Fig. 1) by the method of simultaneously fitting complementary spectra.

Simfit	Absorbers	χ^2	f_s	t_M^a	Γ_s^b	$\Gamma_s+\Gamma_a^b$	δ^b	e^2qQ^b
$\Gamma_a=\Gamma_s$	1,3,5,7,9	1.71	0.4446 ±.0014	4.605 ±.070	0.9503 ---	1.9005 ±.0070	0.9674 ±.0043	-5.07 ±.12
	2,4,6,8	2.43	0.4543 ±.0013	5.086 ±.081	0.9637 ---	1.9275 ±.0068	0.9616 ±.0044	-5.35 ±.11
$e^2qQ < 0$	1,3,5,7,9	1.58	0.4814 ±.0064	3.980 ±.148	1.264 ±.046	2.073 ±.029	0.9595 ±.0052	-4.83 ±.14
	2,4,6,8	2.33	0.4894 ±.0071	4.113 ±.191	1.276 ±.053	2.050 ±.030	0.9458 ±.0059	-5.38 ±.12
	Average	1.96	0.4854 ±.0048 (.98%)	4.047 ±.121 (3.0%)	1.270 ±.035 (2.8%)	2.062 ±.021 (1.01%)	0.9527 ±.0039 (.41%)	-5.11 ±.09 (1.8%)
$e^2qQ > 0$	1,3,5,7,9	1.46	0.4782 ±.0062	3.821 ±.145	1.253 ±.046	2.029 ±.029	1.0452 ±.0053	+5.21 ±.13
	2,4,6,8	1.69	0.4822 ±.0065	3.669 ±.177	1.244 ±.049	1.931 ±.028	1.0717 ±.0064	+6.26 ±.11
	Average	1.58	0.4802 ±.0045 (.94%)	3.745 ±.089 (2.4%)	1.249 ±.034 (2.7%)	1.980 ±.020 (1.02%)	1.0585 ±.0042 (.40%)	+5.74 ±.08 (1.4%)

a In mgEu/cm^2

b In mm/sec

parameters than either Γ_a or Γ_s. We obtained a reasonable convergence and refit the spectra, gradually relaxing the constraints on f_s, t_M, and the ratio of Γ_a to Γ_s. Curiously, both the values for χ^2 and the internal consistency between the set of 5 and the set of 4 improved. The final stages in this process are shown in Table 3. When all constraints were removed, the fits converged more rapidly than any of the previous fits, and the internal consistency, shown graphically in Fig. 4, continued to improve, in spite of the remarkably small statistical errors in the parameters. Finally, this last fit was repeated with $e^2qQ > 0$; the convergence was again rapid, and, compared to the previous fits with $e^2qQ < 0$, the χ^2 values decreased for both odd and even sets of data. Unfortunately, as Fig. 4 demonstrates, the consistency for $e^2qQ > 0$ was somewhat poorer, so that a choice between negative and positive values of e^2qQ is not possible with this work.

Let us study Fig. 4 in detail. The 0% lines represent the average values quoted in Table 3. The values obtained by fitting the odd and the even sets of data are indicated in the figure with the symbols, "o" and "e", respectively, as % deviations from the average, while the statistical errors in the parameters are plotted in the standard "error-bar" notation as % of the parameter values. For $e^2qQ < 0$, we see that every variable parameter from the odd set, except for e^2qQ itself, agrees with its partner in the even set to within or better than its statistical error. The consistency for $e^2qQ > 0$ is somewhat less impressive, but, in view of the small statistical errors involved is still quite satisfactory. In Fig. 4, in addition to the statistical errors in the averages, calculated in the usual manner, for e^2qQ we have indicated with broken error bars sufficiently large "consistency errors" to account for remaining inadequacies of the model.

Finally, in Fig. 5 we have displayed the consistency among the average values of the parameters (as listed in Table 3) for the two signs of e^2qQ. Note that the source parameters, f_s and Γ_s, are completely insensitive to the sign of e^2qQ, and these may immediately be given with maximum precision. The absorber parameters are more sensitive, and their errors must be increased accordingly. The various results of the analysis of data employed in this work are listed in Table 4.

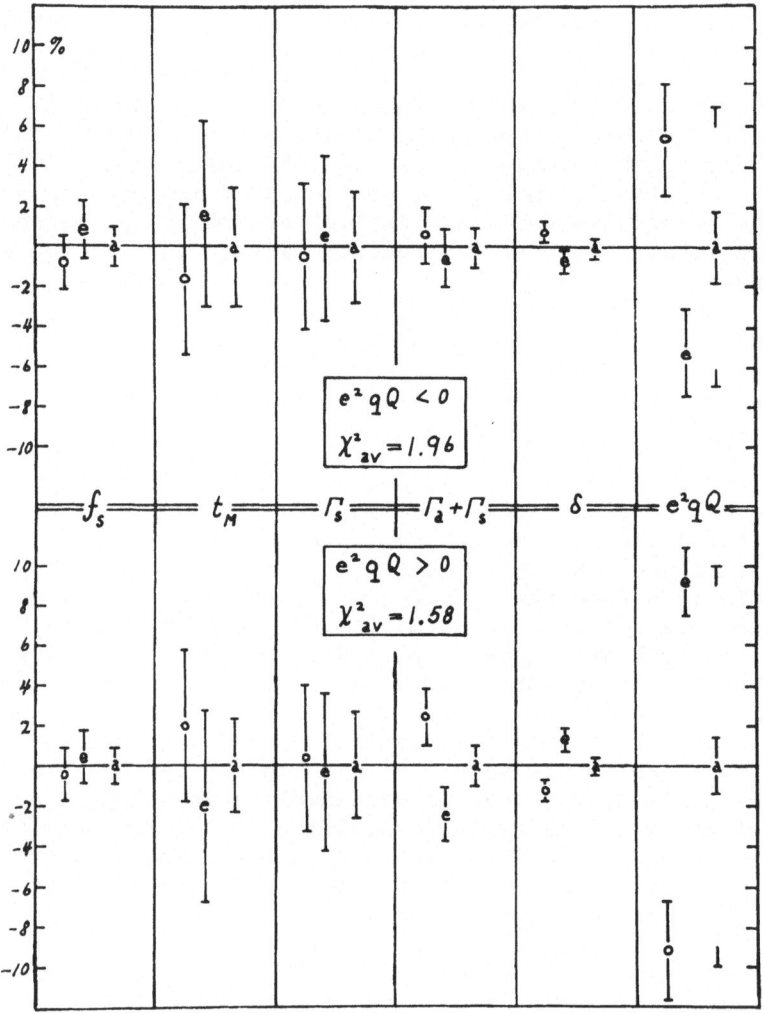

<u>Fig. 4</u>. Graphical representation of the consistency
among the parameter values obtained (See Table 3) by
simultaneous fits to the odd and the even-numbered
spectra. Details of the representation are explained
in the text.

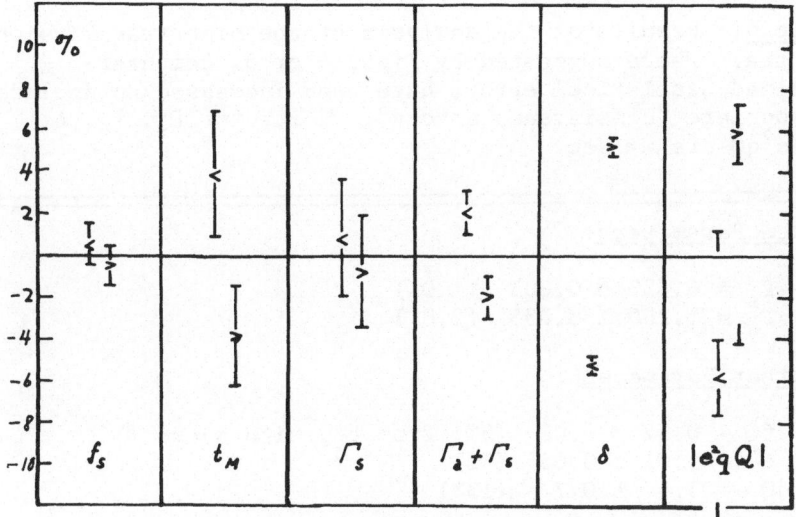

Fig. 5. Consistency among the average values of the parameters (See Table 3) for $e^2qQ < 0$ and $e^2qQ > 0$.

V. DISCUSSION

With this paper we have shown the power of the technique of simultaneously fitting spectra with the transmission integral formulation. Our results show for the first time that it is possible to obtain convergence in a least-squares fit to Mössbauer spectra when all of the transmission integral parameters are varied. We have shown that the method. is sufficiently sensitive to obtain in addition a splitting parameter, e^2qQ, from unresolved spectra. The results themselves are of high precision, while the consistency argument provides a way to assess the reliability of the values obtained for parameters that are not visibly related to properties of the observed spectra. In the present case, the argument of consistency between independent sets of spectra taken for different absorber thicknesses shows that the model used in the final simultaneous fits ($e^2qQ \neq 0$, $\Gamma_a \neq \Gamma_s$) is adequate to explain the data.

Table 4. Results of the analysis of the nine transmission spectra. Where suggested by Figs. 4 or 5, computer-supplied statistical errors have been increased to include appropriate "consistency errors". Units for Γ_s, Γ_a, δ, and e^2qQ are mm/sec.

Source Parameters:

$f_s = 0.483 \pm 0.005$ (1.0%)
$\Gamma_s = 1.260 \pm 0.035$ (2.8%)

Absorber Parameters:

$f_a = 0.57 \pm 0.03$ (5%; from t_M, with $\alpha = 28.5$)
$\delta = 1.01 \pm 0.05$ (5%)
$e^2qQ = 5.4 \pm 0.7$ (13%)

If $e^2qQ < 0$	If $e^2qQ > 0$
$f_a = 0.546 \pm 0.016$ (3.0%)	$f_a = 0.590 \pm 0.014$ (2.4%)
$\Gamma_a = 0.792 \pm 0.024$ (3.0%)	$\Gamma_a = 0.73 \pm 0.03$ (4%)
$\delta = 0.953 \pm 0.004$ (0.4%)	$\delta = 1.059 \pm 0.016$ (1.5%)
$e^2qQ = -5.11 \pm 0.36$ (7%)	$e^2qQ = +5.74 \pm 0.57$ (10%)

Confidence in the model and the method of analysis is further strengthened by the general consistency of our results (Table 4) with earlier work. As an example, $f_a > f_s$ is in line with observations that Eu_2O_3 spectra are generally deeper than EuF_3 spectra [3]. The component width, $\Gamma_a = 0.76 \pm 0.08$ mm/sec, is the smallest yet observed but still somewhat larger than the electronic line width of 0.65 ± 0.02 mm/sec [11]. On the other hand, the source width, $\Gamma_a = 1.26 \pm 0.04$ mm/sec, shows that there is unresolved broadening in anhydrous EuF_3, but the insensitivity of Γ_s to the sign of e^2qQ shows that it is entirely adequate to treat the source spectrum as a broadened Lorentzian. The use of a range of absorber thicknesses has evidently decoupled the precise shape of the source spectrum from the absorber parameters. The isomer shift, δ, is sensitive to the asymmetry in the spectrum that results from quadrupole interaction, in accordance with the prediction of Goodman et al. [1]. Finally, the effective quadrupole interaction, $e^2qQ = -5.11 \pm 0.36$ mm/sec, is in good agreement with the value, $e^2qQ = -4.76 \pm 0.20$ mm/sec, obtained by Glentworth et al. [12] from a Lorentzian analysis with $\eta = 0$ of a single Eu_2O_3 spectrum. This is indeed satisfying, since the negative sign for e^2qQ also gives better consistency between the odd and even-numbered spectra. If we therefore adopt the negative sign, we obtain the results, $f_a = 0.546 \pm 0.016$, $\Gamma_a = 0.792 \pm 0.024$ mm/sec, and $\delta = 0.953 \pm 0.004$ mm/sec.

With the analysis employed in this work we have significantly increased the amount and quality of information that may be derived from unresolved Mössbauer spectra. The parameters of our source, f_s and Γ_s, are now known, and we anticipate that, for future experiments with other absorbers, transmmission spectra may be analyzed individually by fixing these two parameters. We plan to obtain spectra for a set of anhydrous EuF_3 absorbers to better characterize the source spectrum and possibly to provide an independent determination of the internal conversion coefficient, α. Parallel low-temperature [151]Eu and [153]Eu Mössbauer experiments [26] would be of interest in this connection. In fact, the procedure appears so capable of extracting precise information from poorly resolved or weak Mössbauer spectra, that it might help to reverse the trend of neglect in a number of under-exploited Mössbauer transitions.

ACKNOWLEDGMENTS

These studies were supported in part by Grant No. 11,310 from the National Heart and Lung Institute. We wish to thank A. F. Clifford for help in the selection of the source matrix, E. L. Robinson and E. L. Wills for advice in nuclear instrumentation, T. Griswold, who was supported under National Science Foundation Grant GY-10673, for assistance in the data collection, and D. J. Erickson for a number of constructive conversations regarding the theoretical model.

REFERENCES

1. B. A. Goodman, N. N. Greenwood, and G. E. Turner, Chem. Phys. Lett. 5, 181 (1970).

2. R.J.P. Williams, Quarterly Review 24, 331 (1970).

3. G. W. Dulaney and A. F. Clifford, Mössbauer Effect Methodology 5, 65 (1971).

4. R. Gehrke, Nuclear Data document ANCR-1088, p. 392, gives 21.540 ± 0.006 keV.

5. S. Antman, H. Pettersson, Z. Zehlev, and I. Adam, Z. Physik 237, 285 (1970), gives E_γ = 21.529 ± 0.014 keV and α = 28.5.

6. Using δ^2 = 0.00088 ± 0.00007, J. G. Stevens, private communication, calculates α = 28.60 ± 0.15 from R. S. Hager and E. C. Seltzer, Nucl. Data A4 (1968), the error being from the uncertainty in mixing ratio only.

7. E. Kankeleit, Mössbauer Effect Methodology 1, 47 (1965)

8. J. D. Bowman, E. Kankeleit, E. N. Kaufmann, and B. Persson, Nucl. Instr. Methods 50, 13 (1967).

9. ASTM X-Ray Powder Diffraction File, card 12-393.

10. S. Margulies and J. R. Ehrman, Nucl. Instr. Methods 12, 131 (1961).

11. Mössbauer Effect Data Index for 1972, ed. by J. G. Stevens and V. E. Stevens (IFI/Plenum, New York, 1973) p. 146.

12. P. Glentworth, A. L. Nichols, N. R. Large, and R. J. Bullock, J.C.S. Dalton, 546 (1973).

13. G. M. Kalvius, T. E. Katila, and O. V. Lounasmaa, Mössbauer Effect Methodology 5, 231 (1969).

14. H. Frauenfelder, The Mössbauer Effect (W. A. Benjamin, Inc., New York, 1962). Eq. (62), p. 45, has been re-written with the effective absorber thickness, $T_a = t/t_M$.

15. Ref. 14. Eq. (11), p. 7.

16. D. Agresti and M. Belton, manuscript submitted for publication.

17. B. T. Cleveland, Nucl. Instr. Methods 107, 253 (1973).

18. D. J. Erickson, L. D. Roberts, J. W. Burton, and J. O. Thomson, Phys. Rev. B3, 2180 (1971).

19. D. Agresti, M. Bent, and B. Persson, Nucl. Instr. Methods 72, 235 (1969).

20. M. Bent, B. I. Persson, and D. Agresti, Computer Physics Communications 1, 67 (1969).

21. D. Agresti, E. Kankeleit, and B. Persson, Phys. Rev. 155, 1339 (1967).

22. J. W. Burton, ORNL-4743, 105 (1972).

23. J. W. Burton, D. J. Erickson, L. D. Roberts, and D. G. Agresti, to be published.

24. Anthony Ralston, A First Course in Numerical Analysis (McGraw-Hill, New York, 1965).

25. Ref. 14. Eq. (61), p. 45.

26. E. Steichele, S. Hüfner and P. Kienle, Phys. Lett. 21, 220 (1966). These authors study isomer shift ratios in ^{151}Eu and ^{153}Eu.

A MÖSSBAUER BACKSCATTER SPECTROMETER WITH FULL DATA PROCESSING CAPABILITY*

Paul A. Flinn

Carnegie-Mellon University

Pittsburgh, Pennsylvania   15668

INTRODUCTION

The potential usefullness of Mössbauer spectroscopy for a variety of applications in non-destructive analysis and testing has been apparent for more than ten years (1). Nonetheless, the actual practical use of the technique for other than fundamental research purposes has been quite limited. There have been several difficulties: the conventional transmission geometry is unsuitable for most practical applications, while, until recently, backscatter measurements were inefficient and time-consuming; quantitative measurements require computer processing of the raw data, thus introducing additional time delays before final results are available; the measurement and analysis procedures are normally rather complicated and ill adapted to routine use. The purpose of this program was the development of a relatively low cost instrument which could be used

---

* This research was supported by the United States Atomic Energy Commission, Division of Isotopes Research.

by an operator with rather limited training for rapid measurements in a routine manner.

COUNTER DESIGN

The Mössbauer resonance in iron can be observed in the backscatter geometry by detection of gamma rays, X-rays or conversion electrons (2-8). The detection of conversion electrons is unsuitable for the applications envisioned here, since a windowless flow counter must be used. The use of such a counter imposes severe limitations on the sample geometry, since an airtight seal is required; also, a substantial time, is required to purge all the air from the counter before the measurement can start. Both the X-rays and the gamma rays can be detected by a suitable sealed counter; previous work (4,5) has shown that the measurement of the gamma rays is usually preferable to the measurement of the X-rays: although the X-rays are more numerous, the peak to background ratio is so much better for the gamma rays that more information per unit time is obtained by counting them.

The counter must satisfy two requirements: it must intercept as large a fraction of the backscattered radiation as possible, to maximize the useful counting rate; and it must have the highest possible energy resolution, to minimize the background counting rate. The first requirement was satisfied by the toroidal geometry of an earlier design (4), but the departure from the usual cylindrical geometry resulted in a substantial sacrifice of energy resolution. Good resolution in an otherwise unfavorable geometry can be obtained by the use of a cylindrical grid around the anode wire (9). This feature was incorporated into our final design (6), shown in Figure 1. As shown in the figure, massive shielding is required to prevent high energy radiation from the source from entering the counter. A counter of similar design has been reported by Keisch (10). With such counters, the peak counting rate can be more than three times

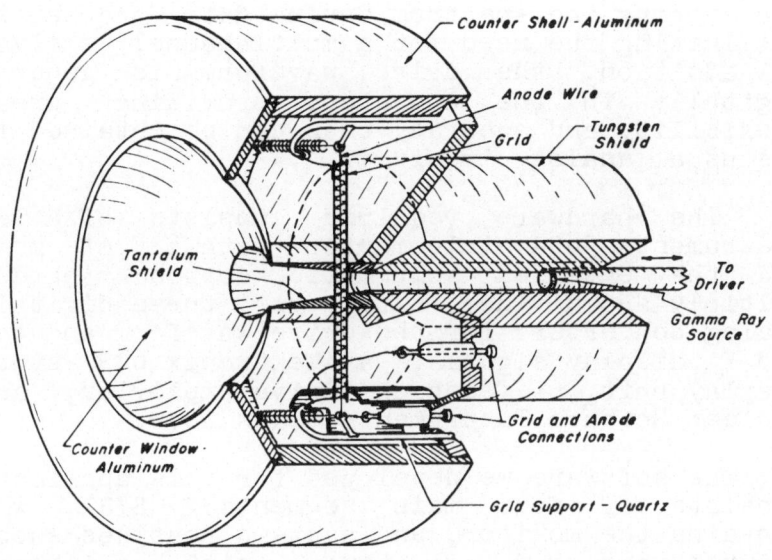

Figure 1. Backscatter Detector.

the background counting rate and a satisfactory
spectrum can be obtained in a few minutes for
materials with a single peak and high iron
content, such as stainless steel. More complex
spectra require somewhat longer times, but a few
hours is adequate for most materials, if a source
strength of about 100 millicuries is used.

SPECTROMETER DESIGN

In most Mössbauer spectrometers, the spectrum
is accumulated in the memory of a multichannel
analyser, additional circuitry synchronized with
the sweep of the analyser memory is used to
generate the drive waveform, and data processing
is carried out in a separate computer, usually a
large one. Since a computer is required in our
instrument for data processing, it is available

for storing the spectrum during data acquisition, eliminating the need for a multichannel analyser. In addition, the drive waveform is generated digitally in the computer, providing greater flexibility and accuracy than can be obtained from the usual analog circuitry.

The hardware required consists of a Texas Instruments 960A minicomputer with 8 K of memory and the extended arithmetic option (hardware multiply and divide), a teletype, three digital to analog converters for the drive waveform and the X and Y display signals, a Tektronix 603 storage display unit, a feedback drive amplifier, and a Nuclear Science Instruments transducer.

The software we developed for this application consists of four main segments: SYS1, which contains the monitor, and systems routines such as input/output and floating point arithmetic; MOSS2, which controls data acquisition; MFIT2, which carries out the data processing, and CDATA, which is the common data storage area. Details concerning these routines are given in Appendix 1. The operation of the instrument is controlled from a teletype; results of the data processing are available both on the teletype and on the cathode ray tube display unit. The data points can be punched out on paper tape if desired.

The drive waveform generated by MOSS2 and converted to analog form by a 12 bit D/A converter is shown in Figure 2. The central portion of the waveform covers the velocity range specified for the spectrum; the high (20 mm/sec) constant velocity portions of the waveform serve two purposes: the counts accumulated during these portions of the cycle are used to determine the background counting rate, and they provide the necessary means for adjusting for retrace when an unsymmetrical velocity range is specified. In the example shown, the velocity range is −2 to +3 mm/sec; since this results in a net positive displacement of the source during the working part of the velocity range, the time spent at −20 mm/sec is automatically lengthed so that the net

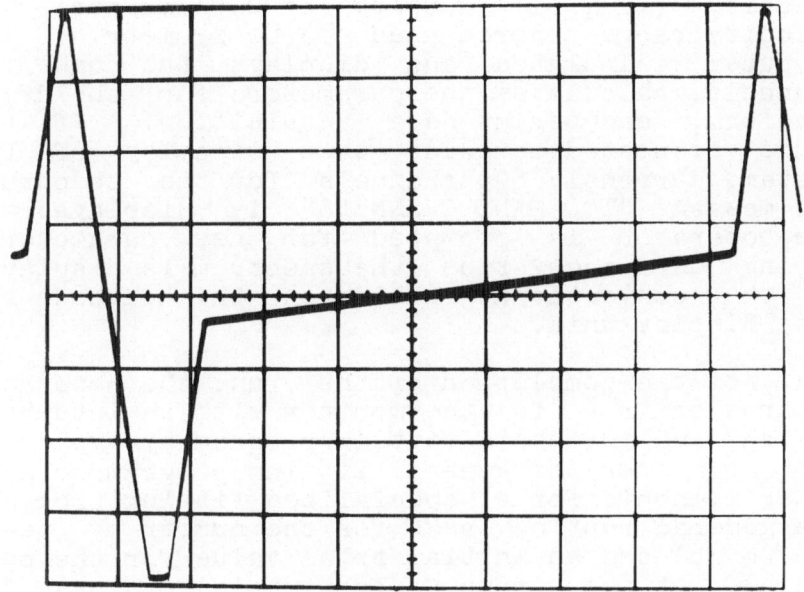

Figure 2. Velocity waveform (-2 to +3 mm/sec).

displacement per cycle (the area of the curve for one cycle) remains zero. The drive amplifier is DC coupled throughout, so that the reference signal must have no DC component. The time per channel is adjusted by the computer according to the number of channels used so that the fundamental frequency is about 18 Hz; e. g., for a 650 channel spectrum, about 70 microseconds per channel. With this frequency of operation, the displacement of the source is small enough to make "parabolic distortion" negligible.

SPECTROMETER OPERATION

To begin a run, assuming the monitor is in control, the operator types NR (new run). This invokes MOSS2, which provides a run number and asks the operator for a label, then for starting velocity (VS), velocity increment (DV), and final

velocity (VF), all in units of 0.01 mm/sec. The
velocity range covered need not be symmetric. The
computer calculates and displays the number of
channels, calculates the parameters for the drive
waveform, and begins data acquisition. If the
range requested exceeds the capacity of the
system, currently 800 channels for the spectrum,
the message "TOO MANY CHANNELS" is displayed and
the operator is prompted for new parameters.
During data acquistion the spectrum is displayed
in the conventional fashion on the cathode ray
tube display unit.

At the completion of the run, the operator
returns control to the monitor with an interrupt
switch, and invokes a fitting program by typing FR
(fit run) for the general fitting program or some
other memnonic for a specialized fitting program.
The general routine asks for the number of peaks
(up to 8) and an initial trial value for the peak
width. It then prompts for initial values for
individual peak position, amplitude and width. If
the operator wishes to supply none, or only some,
of the initial trial values for the peak
parameters, he can invoke a search procedure at
any point by a carriage return instead of a peak
position entry. The search procedure operates by
stripping each peak from the spectrum, placing the
next peak at the highest point in the remaining
spectrum, and continuing until the specified
number of peaks have been located. For the vast
majority of spectra the search procedure works
satisfactorily, and it is unnecessary to enter
parameters for individual peaks.

After the initial parameters have been
determined, the computer carries out an iterative
least squares refinement of the parameters. The
procedure used is discussed in detail in Appendix
2. After each iteration the experimental
spectrum, the calculated spectrum, and the
parameters are displayed on the cathode ray tube.
The calculation is terminated when no further
reduction in CHISQ is obtained with iteration. The
time required for a complete calculation depends
on the number of parameters and on the number of

```
-NR
 RUN   00439
 ENTER LABEL
MACKAY IRON - 4 X .001" 11/2/73
   VS? -650.   DV? 2.   VF? 650.
00650    CHANNELS
-FR BACKGROUND = 05976
NUMBER OF PEAKS?   6.PEAK WIDTH?   35.
PEAK POSITION?
```

POSITION	AMPLITUDE	WIDTH	AREA
-00527	04136	00035	00243
-00299	03447	00032	00185
-00073	01844	00032	00099
00098	01869	00032	00100
00324	03482	00032	00187
00552	04165	00035	00245

```
CHISQ= 00929
```

Figure 3. Typical teletype listing for a run.

data points; some typical times are about 10 seconds for a single peak (three parameter), 200 channel spectrum, and about 5 1/2 minutes for a spectrum with six independent peaks (18 parameters), and 650 channels.

An example of the teletype listing of a run is shown in Figure 3 and the final cathode ray tube display is shown in Figure 4. The background is not a fitting parameter, but is the average number of counts per channel for the high velocity channels. The number of cycles is proportional to the length of the run. The peak positions and peak widths are given in units of 0.01 mm/sec, the amplitudes are given in total counts above background, the "area" is obtained from the product of the peak amplitude and peak width, divided by the number of cycles. The area, therefore, is independent of the length of the run, except for statistical fluctuations.

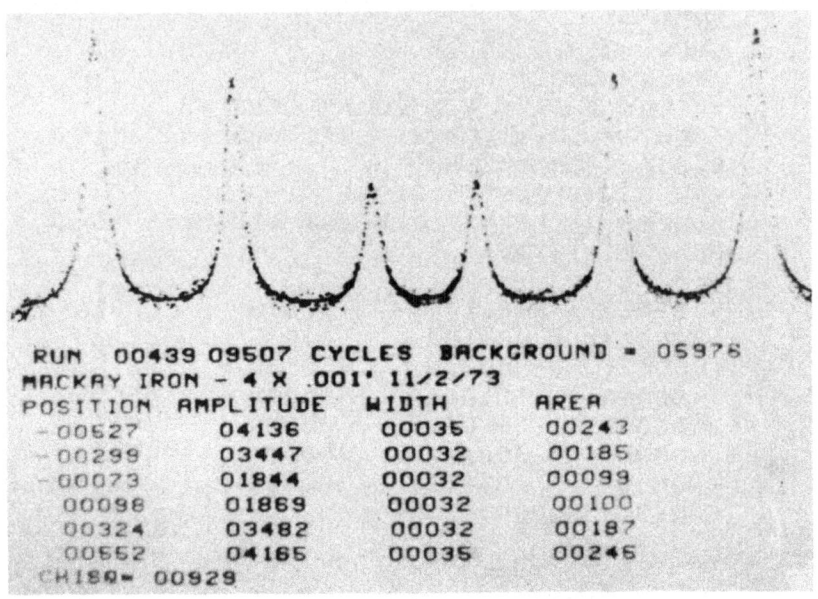

Figure 4.   CRT display of the results of a run.

After the completion of the fitting procedure, the operator can obtain a paper tape of the data points by typing OR (output run). The run can be continued by typing CR (continue run). A new run is started by typing NR. Two additional commands are available for use in special circumstances. One is the command D2 (divide by two), which divides all of the data, both regular spectrum and background, by two. This is necessary when the number of counts is large enough to result in a one in the first bit position, which is the sign bit for certain operations. The other command is used if the spectrometer is operated in the transmission geometry, rather than backscatter geometry; The command is CT (convert to transmission); it causes the contents of each channel to be subtracted from twice the background, so that the dips of the transmission spectrum are converted into peaks, and the spectrum can be processed in the same way as a backscatter spectrum.

APPENDIX 1.

SPECIAL SOFTWARE

The special software written for the 960 A for this application consists of SYS1, the system routine set, which requires 2128 words of storage; MOSS2, the data acquisition program, 492 words; MFIT2, the data processing program, 1790 words; and CDATA, the common data storage block, 3328 words. All together, 7738 of the total of 8192 storage locations are used.

SYS1 is made up of MONIT, the system monitor, 736 words; INOUT, the routines for teletype input and output, and for CRT display, 864 words; FARITH, the subroutines for floating point arithmetic, 400 words; and ARITH, a simple arithmetic program, used primarily to test FARITH, 128 words. The commands available in MONIT include the normal operating commands: NR (new run), CR (continue run), FR (fit run), OR (output run), D2 (divide by two), and CT (convert to transmission). These are discussed in the body of the text. In addition there are system commands: LD (load) which invokes the bootstrap loader; DP (dump) which provides a hexadecimal dump of any specified portion of memory; PH (patch), which permits words to be entered from the teletype in hexadecimal form into any specified memory locations; DG (debug) which permits setting a debug trap at any specified memory location; GO, which causes resumption of program execution after a debug trap has occurred; CC (create command), which provides for the creation of new commands, AH, which causes entry into ARITH; and ??, which lists the existing commands.

MOSS2, which controls the data acquisition, begins with a section which increments and prints the run number, then prompts the operator for run parameters. A zero or negative value for velocity increment is rejected. The number of data

channels needed, (VF-VS)/DV, is calculated, and if
it exceeds 800, the operator is asked for new
parameters. Next the number of extra channels at
high positive or negative velocity needed for
retrace is calculated; 224 channels are available
for the portion of the cycle outside the working
range; this is normally sufficient, but in the
case of a highly unsymmetrical velocity range it
may not be. Then the message: TOO MANY CHANNELS
IN RETRACE is printed and the operator is prompted
for new range parameters. When an acceptable
velocity range is entered, the storage area for
the data is cleared and the data acquisition loop
is entered.

The data acquisition loop is actually a nest
of three loops. The outer, or sequence loop steps
through the portions of the drive cycle: the
negative acceleration from zero to -20 mm/sec, the
constant velocity at -20 mm/sec, acceleration to
the starting velocity, the working portion of the
cycle, then accleration to +20 mm/sec, constant
velocity at 20 mm/sec, and finally negative
acceleration back to zero velocity and the start
of a new cycle. The velocity increment for each
portion of the cycle is selected according to the
index of this loop. The next loop contains the
following steps: increment the velocity register,
clear the external counter in which the gamma ray
counts are accumulated, enter an inner timing loop
for the proper dwell time per channel, read the
external counter, add the new counts to the number
in memory for that channel, enter the channel
number in the register for the X axis D-A
converter, and the number of counts in the
register for the Y axis D-A converter, increment
the channel counter, and continue. At the
beginning of each cycle (the start of the sequence
loop) a panel switch is tested to permit an
interrupt of the run and transfer to monitor.

The module MOSS2 also contains a section
OUTDB, which uses the routines of SYS1 for
printing, and, if desired, punching on paper tape,
the number of counts in each data channel. This
routine is invoked by the command OR.

MFIT2 carries out the general data fitting procedure. First, the background is calculated by averaging the number of counts in the high velocity (+ and - 20 mm/sec) channels, next PARIN prompts the operator for the starting parameters; SEARCH, if invoked carries out a spectrum stripping procedure to obtain initial values for peak positions and amplitudes; then DERIV evaluates the derivatives with respect to each of the parameters at each data point, and combines these to form the matrix of the normal equations, and also calculates the error vector, the vector whose components are the differences between the calculated and observed number of counts at each velocity. The section MINV inverts the matrix of the normal equations by the Gauss-Jordan method; ITER uses the inverted matrix to obtain the solution vector, whose components are the corrections to the parameter values, and then corrects the parameters. A section DRES uses the routines of SYS1 to display the results of the calculation on the CRT display unit. The main memory requirements of MFIT2 are due to DERIV, which requires about 900 words, and MINV, which uses 490. Most of the computing time is spent in DERIV.

CDATA, the common data storage area, contains almost all the numeric data used in the operation of the instrument. Most of the space is occupied by the integer arrays CH1 (1024 words), used for storing the experimental spectrum, and Y (800 words), used for storing the calculated spectrum, and the floating point array WW (1152 words), used for storing the 24*24 matrix of the normal equations. Each floating point number requires two 16 bit words.

APPENDIX 2.

LEAST SQUARES FITTING OF MOSSBAUER DATA

A variety of techniques is available for carrying out least squares fitting of a nonlinear function (11), but most programs for Mössbauer data processing have used either the Gauss—Newton procedure, or a method due to Davidon and Powell (12). The latter method has the extremely attractive feature that it requires only that the fitting function be generated by a user supplied subroutine; analytic evaluation of derivatives by the user is not required. There are, however, two disadvantages to the method which appear to make it unsuitable for use in a minicomputer: it requires a large amount of memory space, and it is relatively slow in execution. The Gauss—Newton procedure requires considerably greater programming effort for each new type of spectrum, but the memory needed is modest, and execution is quite rapid.

The necessary analytic evaluation of the derivatives of the spectrum function with respect to the parameters is readily carried out for the case where the spectrum is taken to be a sum of peaks of Lorentz shape, with the position, amplitude, and width of each peak represented by a separate parameter. This approach has been widely used for the processing of Mössbauer data, and is satisfactory for most purposes. It is the basis of the normal fitting procedure used in this instrument. In some circumstances, however, it is preferable to carry out a fitting procedure which requires fewer parameters; this is possible by incorporating into the calculation some of the physical restrictions known to be present in a particular case. If, for example, the material under study is known to be a simple magnetic material with no quadrupole interaction the only parameters we should need for a complete description of the spectrum are the magnetic field, the isomer shift, and the amplitudes and widths of symmetric pairs of peaks. This is a total of eight parameters, rather than the eighteen required if the spectrum is treated as a sum of six separate Lorentzians. This reduction in the number of parameters results in a considerable saving in the time required for fitting the data. It also makes it possible to fit a spectrum from a material containing two magnetic phases, requiring 16 parameters. In order to fit such a spectrum with independent peaks it would be necessary to use 36 parameters, which exceeds the limit of 24 imposed by the present memory size. More parameters could, of course, be accomodated by providing additional memory, but at additional expense.

The additional programming required to treat special cases is not excessive for most commonly encountered spectra, which can be represented satisfactorily by a sum of Lorentzians, and the natural parameters of the system are related to the parameters of the Lorentzians by linear transformation. The derivatives with respect to the natural parameters are then simply linear combinations of the derivatives with respect to the parameters of the Lorentzians. We can, therefore, accomodate a variety of specialized spectra in a relatively simple way: all we need for each type of spectrum is a short subroutine which forms the appropriate linear combinations of the derivatives which are calculated in the general fitting routine, an input subroutine which prompts for the necessary starting parameters, and an output routine which formats the results properly for display.

REFERENCES

1. L. M. Epstein, "Some Applications of the Mössbauer Effect" in Proceedings of the Symposium on Physics and Non-destructive Testing, Southwest Research Institute, San Antonio (1963)
2. R. L. Collins, Möss. Effect Methodology 4, 129 (1968)
3. K. R. Swanson and J. J. Spijkerman, J. Appl. Phys. 41, 3155 (1970)
4. H. K. Chow, R. F. Weise, and P. A. Flinn, "Mössbauer Effect Spectrometry for Analysis of Iron Compounds" USAEC Report NSEC-4023-1 (1969)
5. B. Keisch, Nucl. Inst. Meth. 104, 237 (1972)
6. P. A. Flinn in "Isotopes Development Program Research and Development: 1972" USAEC Report WASH 1220 p 92 (1973)
7. R. N. Ord, Appl. Phys. Letters 15, (1969)

8. R. L. Collins, R. A. Mazak, and C. M. Yagnik,
Möss. Effect Methodology 8, 191 (1973)
9. R. J. Norman, Austr. J. Phys. 8, 419 (1955)
10. B. Keisch, Mössbauer Effect Spectroscopy
without   Sampling: Application   to Art   and
Archeology", Symposium on Archeological Chemistry
at 165th meeting of the ACS; Advances in Chemistry
(in press)
11. J. E. Dennis, Jr. in "Numerical Solution of
Systems of Nonlinear Algebraic Equations" , G. D.
Byrne and C. A. Hall ed. (Academic Press, New
York, 1973) p. 157.
12. J. D. Powell, Computer Journal 7, 303 (1965)

A METHOD FOR DEPTH SELECTIVE ME-SPECTROSCOPY

U. Bäverstam, C. Bohm, T. Ekdahl,
D. Liljequist and B. Ringström

Institute of Physics, University of Stockholm
Stockholm, Sweden

INTRODUCTION

Lately there has been an increasing interest in ME-techniques involving the detection of resonance scattered conversion electrons. This is mainly due to the inherent possibilities of surface studies. As pointed out, for example, in a number of articles by Spijkerman et al.(1), the detection of conversion electrons in the iron-57 case will limit the surface examined to the outermost hundreds of Ångströms, if metallic iron is used as the absorber. Several papers involving the use of this technique in metallurgical connections have been published recently (2). They are all based on a "depth integral" approach, i.e. all electrons that are scattered are detected regardless of what depth they originate from.

As early as in 1969 Bonchev et al.(3), however, published a paper suggesting the use of a "depth differential" technique, based on the energy loss of electrons penetrating matter. In their paper some experimental results from measurements on tin were presented. In spite of the fact that this paper is commonly referred to, their ideas seem not to have been further elaborated. This is probably due to the experimental problems concerning such measurements, which will be shown later in this article.

The method for performing depth selective ME-measure-

ments that is presented in this paper is based on the same
main ideas as those of Bonchev et al., i.e. the energy loss
of the detected resonance scattered conversion electrons
(or Auger electrons) is used as a measure of their penetration depth in the scatterer. The energy loss of electrons
is, though, a statistical process, and any method using
this energy loss must be based on a knowledge of the energy
loss versus penetration depth distributions of electrons.
This paper begins with a presentation of a method to determine these distributions.

THE ENERGY LOSS DISTRIBUTIONS

The specific energy loss of electrons has been treated,
ever since the discovery of the electron, by a very large
number of authors and in numerous ways. Most works treat,
however, electrons which have energies that are large compared to the atomic binding energies, or at least some ten
or twenty keV. Unfortunately the energies released in ME-
transitions, and especially that in iron-57, are smaller
than that. In fact, there seems to be no theory at present
capable of predicting the energy loss distributions of
electrons of 5 - 10 keV with an acceptable degree of accuracy. Thus, it is necessary to rely on experimental determinations.

Such determinations are, in principle, easily performed.
Monoenergetic electrons are allowed to pass thin layers of
the material to be investigated, and the energy distribution
of the transmitted electrons is determined by means of some
electron spectrometer. A number of layers, with well known
thicknesses in the regions of interest, i.e. from about
50 Ångströms and up to some thousand have to be used. This
procedure is, however, not so easily performed. It is difficult to make thin, homogenous films with well known thicknesses, especially for certain substances. Moreover, the
films created do not necessarily have the same qualities
as "bulk". To overcome these difficulties, a method has
been developed to determine the energy loss distribution
of electrons, after their passage of an arbitrary distance
through a given material. This method has given excellent
results (4, 5). It may be described as a kind of interpolation, using one measured electron energy loss distribution (corresponding to the passage through a rather thick
layer) and the line shape of the spectrometer to obtain

energy loss distributions corresponding to thinner layers.

The principle is the following. A beam of electrons of energy E_O passes a very thin layer of some material. After the passage the electrons will no longer be monoenergetic, but have an energy loss distribution which may be called $f(E_O,dE,x)$, where x is the thickness of the layer. When the electrons are allowed to pass through two layers identical to the first one, they will, of course, have a broader distribution. If $dE \ll E_O$, then the new distribution is found from a convolution of the distribution $f(E_O,dE,x)$ with itself, i.e. $F(E_O,dE,2x) = f \boxtimes f$ where the sign $\boxtimes$ indicates a convolution in the energy domain. Now, if a third layer is added, the electrons appear after the passage of this with an energy distribution given by $F(E_O,dE,3x) = f \boxtimes f \boxtimes f$ and, in general, if the electrons pass a layer of thickness n.x their energy distribution can be given by

$$F(E_O,dE,n \cdot x) = f \boxtimes f \boxtimes f \boxtimes \ldots \boxtimes f \text{ (n times)} \qquad (1)$$

The above reasoning is strictly one-dimensional, while the actual experimental situation is three-dimensional. The method is, however, still applicable as is shown in the comparison with experimental data.

Now, if we in a spectrometer measures electrons that have passed a layer of thickness t, the recorded spectrum $Y(dE,t)$ is the distribution $F(E_O,dE,t)$ modified by the spectrometerfunction $P(E,dE)$,

$$Y(dE,t) = \int_0^{E_O} F(E_O,dE',t) \, P(E_O-dE',dE-dE') \, d(dE')$$

If P is independent of energy, the modification reduces to a convolution of F with P. The layer may be regarded as composed of n thinner layers, each of the thickness $x = t/n$ and with a distribution function $f(E_O,dE,t/n)$. Thus, if the distribution $g(dE)$ is found, which when convolved with itself (n-1) times will reproduce $F(E_O,dE,t)$, then $g(dE)$ is an approximation of $f(E_O,dE,x)$.

$g(dE)$ is obtained in the following way: A "reasonable" analytical expression involving a number of parameters, $g'(dE,\theta)$ is chosen. By an n-fold convolution of P with g' a distribution $G(dE,\theta)$ is obtained. The parameters and consequently $g(dE) = g'(dE,\theta=\theta')$ are determined by fitting $G(dE,\theta)$ to the measured distribution $Y(dE,t)$. Finding a

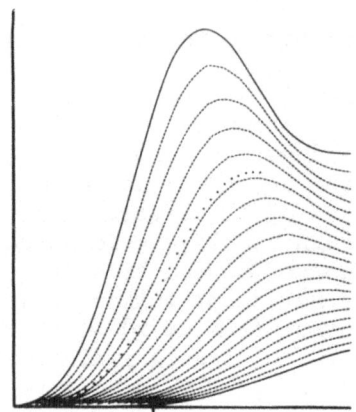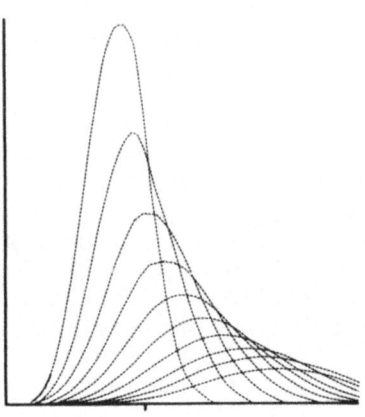

Figure 1. Two examples of "convolution interpolations"; the left example for an energy resolution of about 8 %, the right for about 4 %. The uppermost lines correspond to unscattered electrons, whereas the other full lines correspond to 7.2 keV electrons scattered through a 2570 Å thick foil of iron. The dashed lines in between are calculated from the experimental ones.

"reasonable" expression is, to a certain extent, a matter of trial and error. Once this approximate $f(E_0,dE,x)$ is obtained any $F(E_0,dE,m \cdot x)$ is easily found for $m < n$ simply by an $(m-1)$ fold convolution. If the thickness x of the sublayers is small enough, distributions corresponding to thicknesses between the calculated ones are found by conventional interpolations. Using $m > n$ makes it possible to extrapolate energy loss distributions for layers slightly thicker than the measured one.

In fig. 1 an example of a convolution interpolation as described above is shown, taken from ref. 4. The original energy of the electrons is 7.2 keV, and they are scattered through iron. In spite of the very large difference between the spectrometer function and the measured energy distribution, the calculated intermediary distributions were all found to be in very good agreement with the experimental distributions from a layer of corresponding thickness, as long as $dE \ll E_0$. From a set of calculated distributions it is now easy to find what can be called weighting functions $w(dE,x)$ which show the detection probability of the electrons

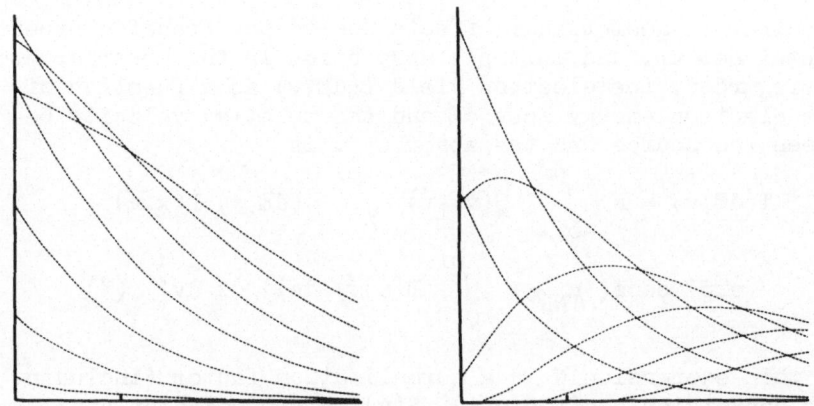

Figure 2. Weighting functions, i.e. the detection probabilities of electrons originating from different depths in a
source, for different spectrometer settings. The bar on the
axis indicates the depth 500 Å. The curves are calculated
from fig.1.

originating from the depth x if the spectrometer is set to
a given energy. The weighting functions for different energies corresponding to the distributions in fig.1 are shown
in fig.2.

If a source of electrons, for example a radioactive
source or a scatterer of resonance radiation, which emits
$P(x)dx$ electrons in a solid angle of 2π from a depth
layer x·dx, is analyzed in a spectrometer set to an energy
loss of dE, the number of recorded electrons $T(dE)$ is given
by

$$T(dE) = \int_{0}^{\infty} P(x)\ w(dE,x)\ dx \qquad (2)$$

This relation is important in the following discussions.

APPLICATION TO MÖSSBAUER SPECTROSCOPY

The knowledge of the electrons probabilities $w(dE,x)$
makes it possible to obtain a semi-analytical expression
for the energy discriminated conversion electron Mössbauer

spectra. If geometrical effects due to the specific experimental set-up, and multiple absorbtion in the absorber are
disregarded, the electron yield T(dE,v) as a function of
the electron energy loss dE and the relative velocity between the source and the absorber v is

$$T(dE,v) = N \int_{-\infty}^{\infty} S(v -\acute{v}) \int_{0}^{\infty} w(dE,x) \, D(x,\acute{v})$$

$$\exp(-\sec\phi(\, \mu_{NR}x + \int_{0}^{\infty} D(s,\acute{v}) \, ds)) \, dx \, dv' \quad (3)$$

In this expression N is a normalization factor (including
the conversion coefficient), S(v) is the emission spectrum
of the source, μ_{NR} the non-resonant absorbtion coefficient,
and ϕ is the angle of incidence of the gamma rays. D(x,v)
is the absorbtion spectra for the absorber at the depth x.
This is, as usual, a sum of Lorenzians with different amplitudes, widths and positions. Several of these parameters
may be depth dependent. They will, in the following, be
denoted by the vector function P(x).

Equation (3) may thus be seen as the formula linking
the "depth integral" spectrum T(dE,v), which is accessible
to direct measurements, to the "depth differential" spectra
D(x,v) which are not. In order to obtain unambigous information about the parameter functions P(x) it has proven
to be necessary to use several energy discriminated spectra
measured at different electron energies.

Two different approaches have been tried. The first
method is easy to handle and gives a visual picture of the
events but it involves approximations and is statistically
inefficient. The second method is statistically efficient
but demands knowledge of, or "inspired guesses" concerning
the depth behaviour of $\underline{P}(x)$. The two methods are comple-
ments rather than alternatives. The first should be used to
obtain starting values for the second.

The Single Channel Method

The first method mentioned above may be described as
a single channel method. Several spectra are measured at
different energy settings in the electron spectrometer, and
the sets of corresponding channel contents are treated

independently of each other. Equation (2) is used:

$$T(dE) = \int_0^\infty P(x)\ w(dE,x)\ dx \qquad dE = e_1,\ e_2,\ \ldots$$

which now is valid for one given channel in the measured spectra. Now a simplifying assumption has to be made concerning the $P(x)$ in the above equation. The most simple calculations are obtained by regarding $P(x)$ as a piecewise constant function, having the values P_i, $i=1,2,\ldots$ between the limits l_i. This is a physically reasonable assumption as, for example, in the case of an oxide layer on the surface of metallic iron, if natural iron was used. The integral above will, in such a case, reduce to the sum

$$T(dE) = \sum_i \left(P_i \int_{l_i}^{l_{i+1}} w(dE,x)\ dx \right) \qquad (4)$$

If here the limits l_i are regarded as known, (4) constitutes a set of linear equations, which may be solved if the number of different spectra measured is larger than the number of P's. If the limits are not known, eq.(4) can still be directly solved, if only the number of measured, different spectra is increased.

Once (4) is solved for all channels in the measured spectra, the calculated P's are analyzed in the conventional way, fitting to sums of Lorenzians. The errors in the fitted parameters will be large, however, because of the poor data reduction performed by (4).

If the P's used in (4) really were constants, then the solutions found should be unbiased. This is never exactly the case, because of the attenuation of the gamma radiation in the absorber. The discrepancies are, however, small if the P's are reasonably constant over the layers chosen, and if the energy settings of the spectrometer are chosen with care. If a separation of the layers using the wrong limits, or otherwise wrong assumptions is tried, then a "mixing" between the different spectra occurs. Peaks from the surrounding layers appear in a calculated spectrum. These peaks may have positive as well as negative intensities, depending on the true physical circumstances. As will be shown below, the appearance of such "mixed in" peaks may be used to determine true layer limits, or to achieve information on relative intensities of peaks etc. It is also clear that if a peak appears in a calculated differential spectrum, even

with a negative intensity, it indicates that somewhere in
the absorber a peak exists at this position in the spectrum.

If there is no physically sound reason for the choice
of piecewise constant P´s, one may of course chose other
analytical expressions. If a polynomial is chosen, the
system of equations is directly solved. However, the introduction of more parameters, as well as the partitioning
in more layers, ruins the statistics effectively, and so it
may sometimes be advantageous to chose expressions such as
exponentials or fermi-functions, where few parameters are
used.

Simultaneous Fitting

The second method consists of a direct fitting of all
measured spectra to eq.(3), which then should be done using
a chi-square minimization. This fitting involves the determination of the parameters in the vector function $\underline{P}(x)$, the
normalization factor N and eventually some parameters in
the line shape of the source $S(v)$. The number of parameters
to fit will easily be very large, especially if there exist
several peaks in the spectra. (The computer program used by
us allows for up to 95 parameters to be fitted simultaneously.) In such a case there exist many local minima in the
chi-square value, and so several precautions have to be made
so that the minimization will converge to the true minimum.
It is advisable to start with a simple approximate technique.
The method described in the preceding section is well suited
to determine the main depth behaviour of the spectra, since
here the "depth differential" spectra are directly visible.
Second, physically reasonable and yet simple expressions
for the functions $P(x)$ should be chosen, for example exponentials or fermi-functions. Next, (3) should be evaluated
for the starting parameters chosen, and compared to the
measured spectra. A "manual" fitting, i.e. a successive evaluation of (3) for different parameter values is recommended,
if possible on a display unit. Finally it is time to use the
computer fitting program. If a fitting is attempted without
proper precautions, the solutions will, in cases as complicated as these, almost inevitably be false.

Some qualities of this method should be pointed out.
All channels in all measured spectra are treated simultaneously, thus the method is much more efficient from a

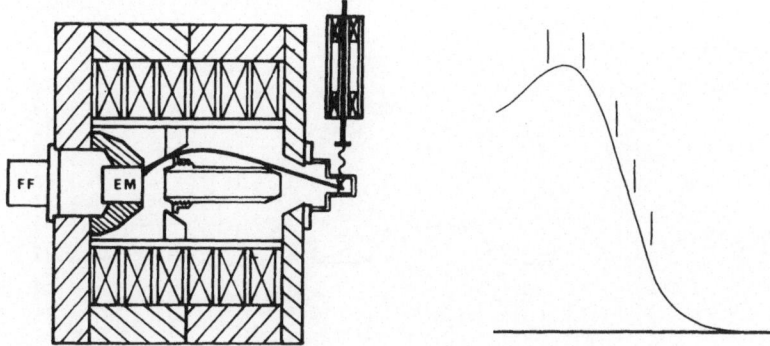

Figure 3. To the left the electron spectrometer and the
transducer are shown. The angle between their axes is near
to 90°, the normal to the absorber makes an angle of about
17° with the spectrometer axis. To the right the electron
line of resonance scattered electrons is shown. Some spectrometer settings used are indicated.

statistical point of view than the earlier described method
is. No knowledge of the cross-section for resonance absorbtion is required (except as a starting value). Thus, the
value achieved in the fitting is a good measure of the validity of the calculations. Moreover, if an expression such
as (3) is used, where the convolution of the source line
shape is performed explicitly, this shape may be altered at
will, or even fitted. Last, but not least, if scattered
electrons are detected, (3) or equivalent expressions are,
as a matter of fact, needed in order to treat the line broadening in a correct way, even in those cases where no depth
variations exist.

In eq.(3) the two simplifying assumptions that geometrical effects (cosine smearing etc.) and multiple absorbtion
can be disregarded, are made. Taking these effects into
account certainly increases the complexity of the formula,
however, there is no principal difference in the fitting
procedure. It is the intention of the authors to incorporate
these effects in the computer program later. (The effects
introduced by the varying source-absorber distance is of
course already taken into account. So are the non-resonant

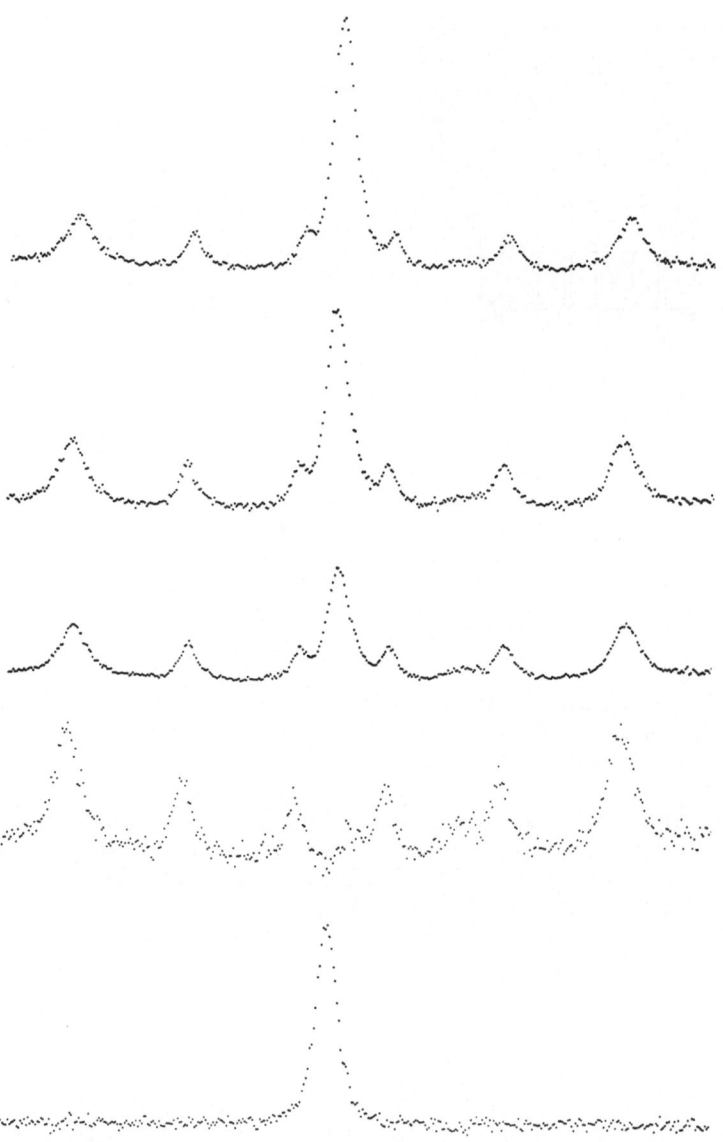

Figure 4. Spectra for a 360 Å layer of iron on stainless steel. From top to bottom: Measured spectra for settings 1, 2 and 3 shown in fig. 3, separated spectrum for the layer 0 - 375 Å, separated spectrum from 375 Å and inward.

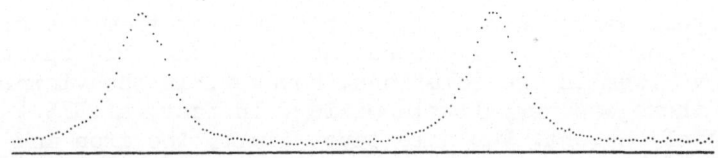

Figure 5. The two middle peaks in an iron spectrum, used for calibration. As seen, these peaks are symmetric.

background and a flat contribution to the background due to electronic noice.)

EXPERIMENTAL DEMONSTRATION

In order to demonstrate the virtue of the depth selective techniques some experimental results are presented. In the measurements, the electrons were selected by means of a magnetic lens spectrometer(5), having an momentum resolution of about 4 %, i.e. an energy resolution of about 8 %. After the passage of the spectrometer the electrons were detected by means of a window-less mesh multiplier (Johnston MM-1). The geometry of the experimental set-up is outlined in fig.3, and described in some detail in ref.(4). The source used was ^{57}Co in Pd-matrix, with an approximate strength of 40 mC. All spectra shown are recorded for about 24 hours.

First the results of a very clear-cut case are shown. A layer of iron (90 % enriched) was electrodeposited on to a foil of stainless steel (NEN 310, 90 % enriched), and the sandwich was then annealed at 300°C for 3 hours in H_2. At this temperature the diffusion rate is so small that the iron layer may be expected to remain iron. This absorber (no 1), as well as the one described below (no 2) were made by NEN, and it was specified from the manufacturer that the thickness of the iron layer was 360 Å. This was calculated from the electrodeposition rate. Three spectra were recorded at the three left spectrometer settings indicated in fig. 3. The maximum number of counts per channel in the steel peaks were respectively 9000, 8000 and 11000 above the background. These spectra were analyzed using the approximation of piecewise constant P's as described in the preceding paragraph.

Two layers were assumed, the limit between them was varied
in steps of 25 Å. As expected the steel and iron spectra were
usually mixed in the solutions, however, in the vicinity of
350 Å there was very little mixing. In fact, at 375 Å it was
completely absent. For this layer limit, the iron and steel
spectra were fully separated, as shown in the two lowest
spectra in fig. 4. The three upper spectra are the measured
ones. Fig. 4 is a convincing demonstration of the potential-
ity of the method. It should once again be pointed out that
the spectra are separated without any assumption concerning
their shapes.

The next example is a more intricate case. The absorber
used here was made identical to the first one, except for
that the annealing temperature used here was 800°C. Spectra
were recorded for all five settings shown in fig.3. First
the same velocity display as that in fig. 4 was used. No
sign of largely splitted peaks were found (the iron should,
in fact, have been wholly transformed into steel in this
annealing), and so a larger display was used. For comparison
a serie of spectra on a pure stainless steel 310 absorber
(no 3) were also recorded. The measurements of the two series
were sandwiched in order to avoid systematic errors. In fig.6
the measured spectra from absorber no 2 are shown. Iron
spectra were also recorded for velocity calibration, and to
test the geometrical effects (possible skewness etc.).

In the analysis a single channel method was first used.
From this it was evident that the small "bump" at the right
slope of the steel peaks was a real peak, situated in the
surface region of the absorber. Next, the peaks from the
measurements on pure steel were simultaneously fitted. It
was found that a sum of at least three Lorenzians had to be
used in order to reproduce the peaks.(The peaks were all
skew, and to achieve a good fitting it was necessary to use
an additional linear parameter in the Lorenzian line shape
expression. In the analysis of an earlier series of measurements on steel, no such term had to be included. This was
probably because of poor statistics, which made the chi-
square value small in spite of a wrong solution (a well known
phenomenon, cf the article by Ruby (6)). The shift of the
steel peaks with depth, reported by the authors in (4), which
was not reproduced here, can certainly be explained by the
neglection of this anomalous line shape, which would exagger-
ate any shifts observed in measured spectra.)

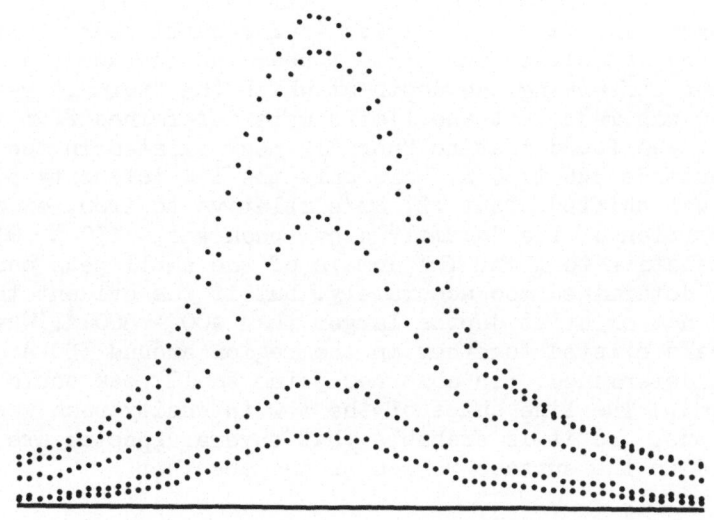

Figure 6. Spectra measured at positions 1, 2, 3, 4 and 5
from left in fig. 3. The absorber (no 2) was stainless steel
with a layer of 360 Å iron on the surface, annealed at 800°C
for 3 hours in H_2. A small extra peak is seen on the right
slope (positive velocity to the right).

The intensities of the peak in all spectra from absorber
3 were well fitted, the calculated absorbtion coefficient being in good agreement with the cross-section for resonance
absorbtion.

In the next step in the analysis of absorber 2, the
same line shape as that of absorber 3 was used as a starting
value, and the five measured spectra (fig.6) were fitted
simultaneously to eq.(3). First the assumption that the
"normal" peak existed right through the absorber was tested.
This was immediately found not to be true. First, the intensities of the peaks from absorber 2 were all found to be
about 40 % smaller than the corresponding intensities from
absorber 3. Moreover, the relative intensities between the
spectra from 2 could not be fitted simultaneously. Thus it
was clear that the "normal" peak did not exist all through
the absorber. Therefore a solution was tried where the peak

was absent in some depth region (and a small peak, shifted
at higher velocities was present near the surface). No assumption concerning the depth at which the "normal" peak
existed was made, but the limits were determined from the
fit. It was found that no "normal" peak existed in the region outside 346 ± 30 Å. Here only the low intensity peak,
which was shifted about .27 mm/s relative to iron, existed.
The position of the "normal" steel peak was -.130 ± .010
mm/s relative to iron. The domain of the small peak could
not be determined too accurately, but it was evident that
it did not exist at depths larger than 400 - 500 Å. Whether
the peaks existed together in the region around 350 Å could
not be determined. (In absorber 3 the small peak could not
be found.) The line shape of the low intensity peak was very
anomalous, and it is probable that several spectra are superimposed in the surface region of the absorber.

STATISTICAL CONSIDERATIONS

The errors in the calculated parameters depend, of
course, on the number of counts in the measured spectra,
and the number of spectra measured. Moreover, they depend
on the different settings of the electron spectrometer, the
partitioning of the total measuring time between the different spectra, and the resolution of the spectrometer.

It is obvious that, for a given total number of counts,
the precision in the results is higher the better the resolution in the spectrometer. If, however, the measuring time
is held constant, the transmission of the spectrometer, or
rather its luminosity, must be taken into account. If, for
a given spectrometer, the relation R/T = constant is assumed,
where R is the resolution and T is the transmission, there
exists an optimal R. If the resolution is too bad, no depth
variations may be determined even if the statistics are
good. If the resolution is very good, the accuracy in the
determination of the weighting functions $w(dE,x)$ becomes
crucial, spoiling the accuracy of the results. It is probable, however, that the region of R is rather large. To
illustrate the effects of the electron spectrometer resolution we show in fig. 7 some computed spectra, corresponding
to the spectrometer resolutions of fig.1. The following
physical properties of the absorber were simulated: The
f-factor, abundance etc. were constant through the absorber.
Three different layers existed, one from 0 to 200 Å, a

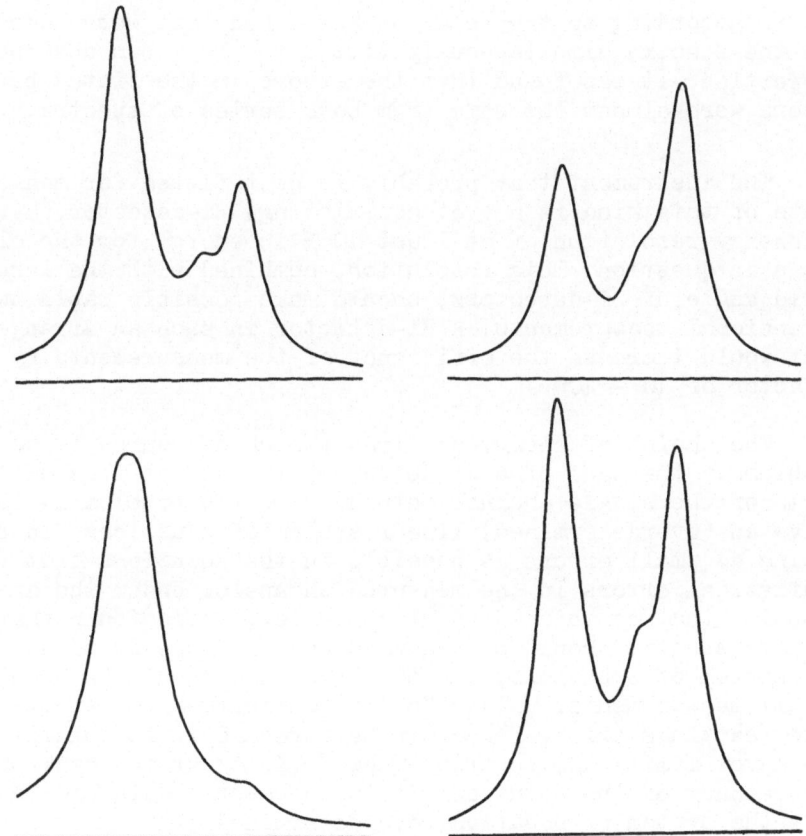

Figure 7. Simulated spectra for an electron spectrometer resolution in energy of 8 % (above) and 4 % (below). The spectrometer settings used are number 1 (left spectra) and number 5 (right) of fig. 3. The peaks emanate from the layers 0 - 200 Å, 200 - 225 Å and 225 Å and inward (from right to left peak in the spectra).

second from 200 to 225 Å and a third from 225 Å and inward in the absorber. One peak existed in each layer, and this was shifted (from the right to the left in fig.7). The absorber was supposed to contain 50 % iron-57.

Five spectra, corresponding to the settings in fig. 3 were generated for each spectrometer resolution. Statistical

errors according to the relation R/T = constant were added,
and the spectra simultaneously fitted to the known absorber
properties. It was found that the errors in the fitted parameters were almost the same from both series of spectra.

The instrument that probably is best fitted for measurements of this kind is a (yet nonexisting) Si-detector, having
an energy resolution of at least 10 % in energy for the elec
trons in question. This resolution, combined with the large
solid angle of Si-detectors, should make possible rapid two-
dimensional measurements. A Si-detector in such an arrangement would increase the efficiency of the measurements by
a factor of 10 - 100.

The choice of energy settings (which of course is no
problem in the case of a Si-detector) is easiest treated in
terms of the single channel method. Here the problem is to
solve an (overdetermined) linear system of equations, introducing as small errors as possible in the solutions from the
statistical errors in the measured channels. Under the assumption that non-biased solutions exist, it is found that
no more spectra should be measured than are needed to solve
the system of equations, and the same time should be spent
in all measurements. Unique "best" combinations of spectrometer settings exist. These can be foresaid if for example
the layer limits are regarded known. (In order to check the
consistency of the solutions it is, however, advisable to
make the system of equations overdetermined.)

A difficulty related to the problem of statistics is
the possibility of singularities of the determinant of the
system of equations. In the vicinity of these, the solutions
are numerically unstable and often misleading.

In the case of simultaneous solutions, the problem is
more difficult. Probably the best way to determine the measuring conditions is by computer simulations. This is, unfortunately, a somewhat expensive way. As a thumbrule one
may use the following: measure at least three different
spectra, distributed around the peak position of the electron line in such a way that the weighting functions are as
"unequal" as possible, and spend the same time in all
measurements. In doing so not too much time is wasted.

CONCLUDING REMARKS

In this paper a method for depth selective ME-measurements has been described. The method has been shown to work excellently in two cases involving iron-steel systems. It seems probable that most applications of the method will be found in the metallurgical field, for example in studying surface behaviour of alloys, diffusion, corrosion etc. The depth sensitivity of the method depends mainly on two things: the energy of the scattered electrons and the qualities of the material investigated. It is evident that iron-57 here, as in so many other cases, is a very suitable isotope. If the energies of the conversion electrons are too high to allow for a good depth selection, it is of course possible to detect Auger electrons instead. Often, however, Auger spectra are rather complicated and high resolving spectrometers are required to achieve unique depth determinations. (If the weighting functions $w(dE,x)$ used were determined solely from experiments, i.e. without any interpolation, then the original energies of the electrons used might as well be different. So, in the case of iron, spectrometer settings on the Auger line at about 5 keV (where a substancial fraction of conversion electrons are recorded) could be used, bringing about a higher counting efficiency and perhaps a slightly increased depth sensitivity.)

In the case where the analyzed absorber is composed of several different substances, like oxide layers on a metal surface, then the weighting functions are not the same through the absorber. This case may be handled by allowing the weighting functions to vary in the fitting to eq.(3). The number of parameters is not increased by this. A more simple approach is possible, though. If the shapes of the weighting functions are unaltered, and only their depth scale is changed (which is approximately the case in many cases), then a scaling only of the results is needed.

This paper has been wholly aimed at depth measurements by means of scattered electrons. It should also be pointed out that there exists a possibility of performing similar measurements by means of scattered gamma- or X-rays. Here the trick is to measure at different angles of incidence. Weighting functions similar to those used by us can then be calculated.

REFERENCES

1. J.J. Spijkerman in An Introduction to Mössbauer Spectroscopy,(ed. L. May, Hilger, London, 1971)

J.H. Terell and J.J. Spijkerman, Appl.Phys.Letters 13
11 (1968)

K.R. Swanson and J.J. Spijkerman, J.Appl.Phys. 41, 3155
(1970)

2. L.H. Bennet and L.J. Swartzendruber, Proc.Int.Conf.Appl.
Mössbauer Effect (Ayelet Hashahar, Israel, 1972)

H. Onodera, H. Yamamoto, H. Watanabe and H. Ebiko, Japan
J.Appl.Phys. 11, 1380 (1972)

H. Leidheiser Jr, G.W. Simmons and E. Kellerman, Intern.
Corrosion Forum (Anaheim, Calif. USA 1973)

3. Zw. Bonchev, A. Jordanov and A. Minkova, Nucl.Instr.and
Meth. 70, 36 (1969)

4. U. Bäverstam, C. Bohm, B. Ringström and T. Ekdahl, Nucl.
Instr.and Meth. 108, 439 (1973)

U. Bäverstam, T. Ekdahl, C. Bohm, B. Ringström, V. Stefáns
son and D. Liljequist, to be published in Nucl.Instr.and
Meth 115

5. T.R. Gerholm, R. Othaz and M.S. El-Nesr, Ark.Fys. 21,
253 (1962)

6. S. Ruby, in Mössbauer Effect Methodology 8, (ed. I.J.
Gruverman and C.W. Seidel, Plenum Press, New York 1973)

CURVE FITTING AND THE TRANSMISSION INTEGRAL: WARNINGS AND SUGGESTIONS

G. K. Shenoy and J. M. Friedt

Laboratoire de Chimie Nucléaire
Centre de Recherches Nucléaires
67200 Strasbourg, France

H. Maletta

Institut für Festkorperforschung der
Kernforschungsanlage, 517 Jülich 1
Federal Republic of Germany

S. L. Ruby[*]

Argonne National Laboratory
Argonne, Illinois 60439

1. INTRODUCTION

Every spectroscopy has developed its own techniques of
spectral analysis and Mössbauer spectroscopy is no exception
to the rule. The earliest work on the detailed spectral
shape of nuclear resonance gamma rays date back to the papers
of Margulies and Ehrman[1] and Ruby and Hicks.[2] In these they
showed the importance of absorber and source thicknesses on
the shape of individual absorption lines. However, most of

---

[*] Partially supported by the U.S. Atomic Energy Commission.

us in the last 12 years have been satisfied to describe
them in their simplest approximation, i.e. by a sum of
Lorentzian lines whose amplitudes, widths, and positions are
determined by the parameters of the hyperfine interactions.
This "thin" approximation neglects the effects due to
saturation by a thick absorber. Occasionally, the inaccuraci
resulting from the use of the simple approximation in the
analysis of partially resolved spectra have been demonstrated
but any enthusiasm for using the transmission integral,
described in the above mentioned references, was curbed by
the necessity for more elaborate computational facilities.

There have been numerous attempts to overcome this
drawback, mainly by expressing the absorption profile in
terms of a fast converging mathematical series. Most of
this work has been done by Heberle, Cleveland and their
collaborators.[3,4] It was centered around the problems of
iron—namely of one, two or six lines. The one line and
two line problems have been dealt with by numerous authors
who presented certain correction schemes to the "thin"
analysis to account for the absorber thickness effects. In
situations involving more than two overlapping lines, or
where the absorption shape is not really based on Lorentzian
shapes at all (e.g. relaxation spectra), and where the
absorber is reasonably thick (as is usually desired to get
adequate data), no convenient, fast procedure has been
generally available. More precisely, transmission integral
programs have generally not been fast enough to run as part
of least-mean-square minimization techniques.

During the last few years, the pressure of more difficu:
measurements and need for better results has pushed several
groups toward improved procedures. The work of the group
centered at ORNL is especially to be noticed.[5,6,7] Another
way to approach the problem was introduced into the literatu
by the group at Hamburg.[8] They use an accurate (albiet slow
convolution of the expected transmission with the flux to
produce a properly calculated Mössbauer spectrum. This is

then given to a minimization routine using the 'thin'
approximation to produce "fitted" values for the parameters.
By comparing the "fitted" with the input values of the
transmission parameters, one is made aware of the possible
problems, and even given information for making plausible
corrections. We call such graphs 'awareness' diagrams, and
will display some below.

Another technique suggested was introduced at the 7th
Methodology Symposium by Ure and Flinn.[9] Their procedure is
to deconvolute the experimental data using the known
properties of the flux and fast Fourier transform algorithms;
the result is (nominally) the absorber's transmission.
After taking its logarithm, the "thin" approximation is no
longer approximate, and the minimization can proceed
unhindered. Their article, and the following talk by Lin
and Preston can be consulted for further information on the
convenience and accuracy of this procedure.

Recently, Ted Cranshaw[10] of Harwell has made a major step
toward a fast transmission integral routine by pointing out
that the flux and transmission need be calculated only once,
and that the counting rate at any velocity is found by
convoluting these fixed results with an appropriate velocity
offset. We have utilized his idea twice— once in a
straightforward way, and second, by combining it with
Gauss-Legendre integration techniques, looking toward even
faster programs.

The purpose of our talk today is to warn you of the
dangers in the "thin" approach for the analysis of the
partially resolved spectra. We shall present some awareness
diagrams described above for a variety of physical and
chemical problems. In addition we will discuss the analysis
of some actual data using the above-mentioned convolution
procedures. For those of you with access to substantial
computational facilities, we offer the Gauss-Legendre as
well as Cranshaw algorithms as Fortran IV programs.

2. THE TRANSMISSION INTEGRAL

We wish to compute the counting rate in the detector as a function of the relative velocity of source and absorber. It is convenient to describe the flux distribution from the source in three parts:

$$\text{FLUX }(E,v) = B + S(1 - f_s) + Sf_s\, F(E,v)$$

$$F(E,v) = \left(\frac{2}{\pi\Gamma_s}\right)\left(1 + \left(\frac{E - E_o - E(v/c)}{\Gamma_s/2}\right)^2\right)^{-1} \tag{1}$$

where B is the background counting rate, S is the integrated signal rate, f_s is the source resonant fraction, E_o is the nominal γ-ray energy, and E is the actual energy. The transmission of the first two terms is practically independent of E and v. For the last term the transmission is

$$\text{TRANS}(E) = \exp\left[-T_a * A(E)\right] \tag{2}$$

where $T_a = n\sigma_o f_a$ is the absorber thickness (since n is the number of resonant nuclei per cm^2, and σ_o is the cross section per nucleus, f_a is the absorber resonant fraction, T_a is dimensionless) and A(E) is the normalized absorption shape, to be discussed below. The transmitted flux then becomes

$$\text{Rate}(v) = \int \text{FLUX}(E,v) * \text{TRANS}(E)\, dE$$

$$= B + S(1 - f_s) + Sf_s \int F(E,v) * \text{TRANS}(E)\, dE$$

$$= R_0\left(1 - f_s \frac{S}{S + B}(1 - TI(v))\right). \tag{3}$$

Note the pertinent properties of the source and detector are all summarized in the number

$$\text{FSB} = f_s \frac{S}{S + B}$$

while the absorber is fully described by its thickness T_A and the normalized absorption shape A(E). The name TI abbreviates 'transmission integral.'

The absorption shape $A(E)$ is determined in standard fashion from the spins of the excited and ground nuclear states, from the hyperfine interaction parameters, the nuclear moments, etc. The only unfamiliar variation is the normalization required in order for T_A to have its conventional meaning. For the simplest case of a single Lorentzian absorber line, A becomes

$$A(E) = \frac{\Gamma_0/\Gamma_a}{1 + \left(\frac{E - E_1}{\Gamma_a/2}\right)^2} = \frac{\Gamma_0/\Gamma_a}{1 + x^2} . \tag{4}$$

Here Γ_0 and Γ_a are the natural and absorber full widths. If there are L lines of equal width, and W_i is the normalized Clebsch-Gordon intensity for the ith line, A becomes:

$$A(E) = \Gamma_0/\Gamma_a \sum_{i=1}^{L} \frac{W_i}{1 + (x_i)^2} . \tag{5}$$

More generally, the absorber shapes are not limited to sums of Lorentzians. Relaxation spectra in particular give rise to more exotic shapes. The important rule for normalizing A is that its integral must always have the same value, namely

$$\int_{-\infty}^{\infty} A(E)\, dE = \Gamma_0 \pi/2. \tag{6}$$

It should be noted that FSB presents a minor problem. If it is known from auxiliary experiments, then it should be used, and T_a will have the desired absolute significance. If unknown, there is a temptation to make it a variable of the fit. In some cases with very thick absorbers, this may be appropriate, but usually good fits will be difficult to obtain. Roughly, FSB and T_a are 'parallel' variables; increasing either stretches the vertical scale in a very similar way, and the computer naturally has trouble separating the two. It is usually better to fix FSB with an approximate value.

In the next section we describe two TI procedures that we have used.

3. CONVOLUTION PROCEDURES

3a. Gauss-Legendre Quadrature Technique

The principal effort in obtaining the exact spectral shape is of performing the numerical integration in Eq. (3) to an accuracy sufficient to describe the experimental situation. This can easily be done by the Gauss-Legendre quadrature technique,[11] which is applied here to every zone into which the region of interaction is divided. The integrand in Eq. (3) is firstly divided into two major regions the inner region and the wings. The inner region is bounded approximately by the usual velocities employed in the experiments, and is subdivided into smaller zones by laying a uniform grid. In the wings, the size of the zones increases with a power law, and the limits of the integration is thus extended up to about 10^5 to 10^6 times the line width. The integral is now evaluated for each zone, using 12 integration steps. The number of grids in the inner region itself varies from 2 to 20 depending on the experimental velocity range and the source and absorber line widths.

Note that the transmission is calculated only once per spectrum, a la Cranshaw. But flux must here be recomputed anew for every experimental velocity. However this program is rather easily adapted to accommodate experiments performed with any drive waveform. In Appendix A we give a complete FORTRAN listing of a program that one can use. The writeup is such that it can easily be adapted to various physical situations.[12,13]

In the program given, the number of integration steps is set equal to 12. The inner region is divided into sub-zones of width equal to the sum of source and absorber linewidths. For test purposes, the number of these subzones has been increased or decreased by the parameter NADSUB. The procedure described here, with variations, has been successfully used in a variety of problems.

3b. The Cranshaw Technique

The main point seen by Cranshaw is that FLUX and TRANS in Eq. (3) need only be calculated once. To produce all the different rates at every velocity, it is always the same

flux shape and the same transmission that needs to be
convoluted— the only change is their relative displacement.
In fact, if we think of FLUX and TRANS as arrays, FLUX(N,M)
and TRANS(N) where N and M are index numbers, then

$$\text{Rate(M)} = \sum_N \text{FLUX(N, M)}*\text{TRANS(N)}. \tag{7}$$

Once FLUX and TRANS have been computed, then each rate at a
new velocity is only some 400 multiplications and sums.
This must be repeated for each point in the spectrum.

There are, of course, many details in achieving this
result and we have attempted to write the program in a way
which makes it quite easy to vary from one problem to another
(See Appendix B). The program has 5 major divisions. The
first is just common blocks and dimension statements, and
will need to be modified in a straightforward way in changing
from one isotope, or Hamiltonian, to another. The second
division, called initialization, is mainly concerned with
getting the latest values of the interaction parameters
ready to go; it will also need changes. The third section,
called transmission, uses the values of the parameters to
calculate the transmission at every energy. It, too, must
be modified for every new problem. The modifications, till
now, are no more or less than is needed using the 'thin'
approximation. The fourth section, which calculates flux,
will rarely be modified, since we nearly always use a simple
Lorentzian shape here. The last and fifth section, called
convolution, should never need changing from one problem to
another. One exception is where the desired spectrum has a
center of symmetry. It is then faster to calculate up to
this center, and get the remainder by folding and interpolation. This has been left out here for simplicity.

An important program choice is called CONVO. For
CONVO = 0, then the program does not proceed beyond the
third section, produces the 'thin' approximation, and
behaves like the rapid routines we had been used to. For
CONVO = 1, the program does the complete transmission
integral properly. Initial fits are made with CONVO = 0
until one understands the problem rather well; finally, fits
with CONVO = 1 will either give better fits or show that the
'thin' approximation was adequate.

4. SOME APPLICATIONS OF THE TRANSMISSION INTEGRAL

4a. Hyperfine Magnetic Interactions

The error introduced in the determinations of either the ratio of the nuclear g-factor of the levels in a Mössbauer transition or the magnetic field at the nucleus through the 'thin' analysis of a hyperfine magnetic spectrum has been discussed at various places.[14,15,16]

In Fig. 1 we reproduce the 'awareness' diagram given by Gerdau et al.[8] for the case of 2 → 0 transition and a magnetic interaction. A large number of rotational states among the rare earth and the actinide nuclei are analogous cases. We notice from Fig. 1 that the error would be some 17% in the determination of the excited state splitting when the splitting is equal to the observable line width for an absorber thickness equal to 10.

A particularly illustrative example on the accurate determination of the ratio of the nuclear g-factor is given by Gerdau et al. for the 81-keV, $\frac{5}{2} \to \frac{1}{2}$ transition in ^{133}Cs. In Fig. 2 we give the experimental data with least-squares fit obtained using both a 'thin' and TI analysis. In Fig. 3 the values of χ^2, the goodness-of-fit parameter, has been plotted as a function of the ratio of the nuclear g-factors. It is interesting to observe that χ^2 goes through a sharp minimum for a particular value of g_{ex}/g_{gd} only when the transmission integral is employed in the spectral analysis.

The binary alloys of the 3d elements have been investigated over many years to try to understand the relative contributions to the hf magnetic field from various causes.[17] Often the Mössbauer results have been found to deviate from corresponding spin-echo measurements.[18] One of the likely reasons[19] for such discrepancies could be the absorber-thickness effect. Another experimental area where absorber saturation effects are likely to be of importance is in the study of the critical region in ordered magnets. As one raises the temperature, the resonance lines approach ever closer to each other and the overall absorption shape becomes much deeper, narrower, and unresolved. Even if the 'thin' approximation were adequate well below T_c, it is unlikely to remain so when near T_c.

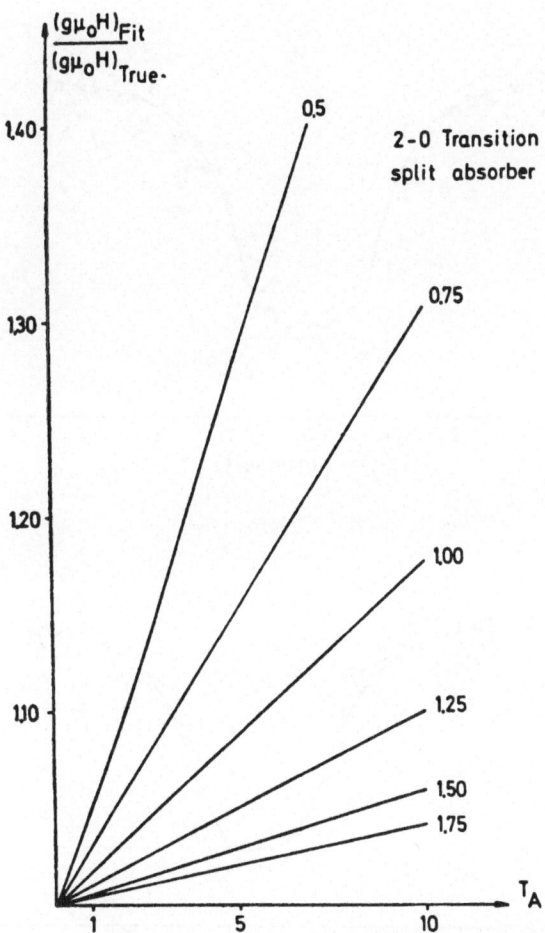

Fig. 1.   Awareness diagram for the magnetic
          hyperfine interaction in the $2 \to 0$
          transitions.   The number on each line
          gives the value of $(g\mu_0 H/\Gamma_0)_{TRUE}$.
          [Ref. 8]

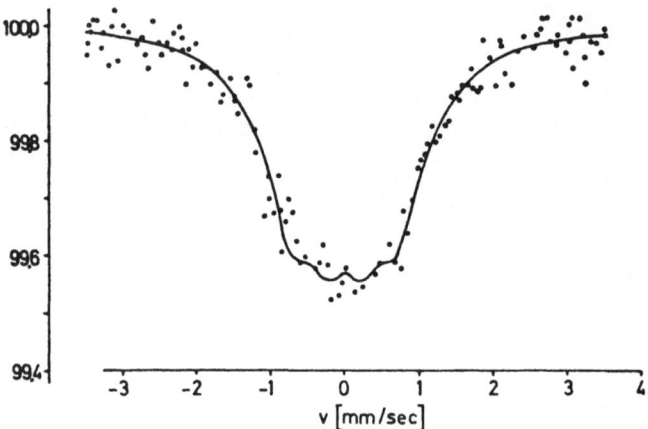

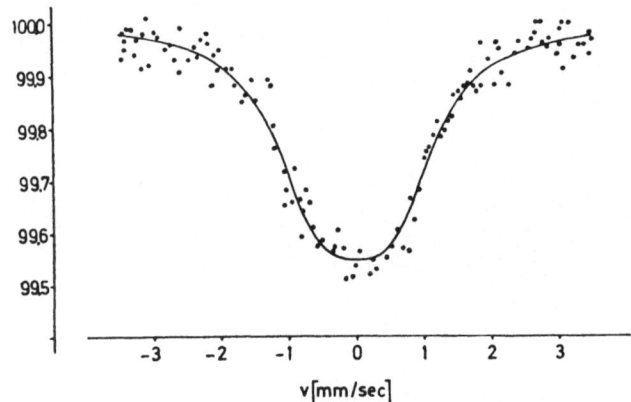

Fig. 2.   The spectrum of ^{133}CsCl in 65 kOe
external field. The solid curves show
results of 'thin' (below) and transmission
integral (above) analysis [Ref. 8].

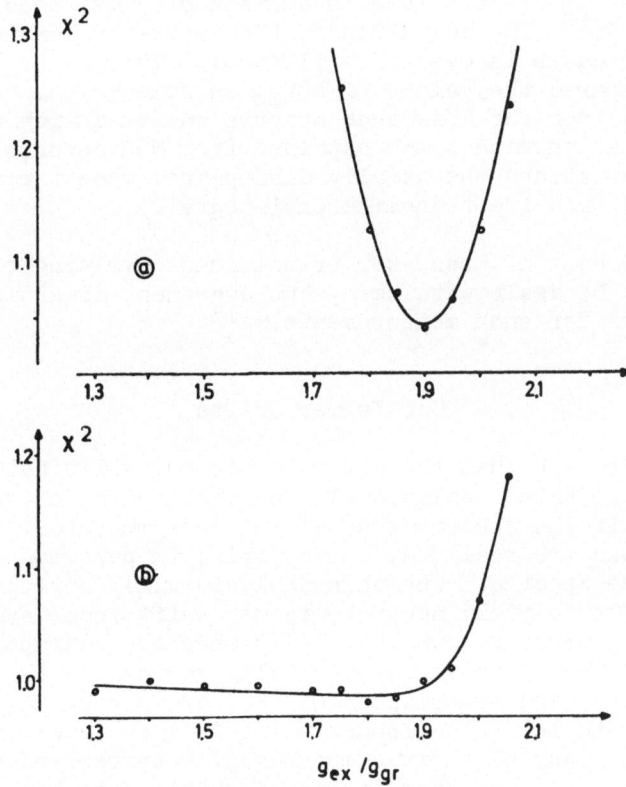

Fig. 3. The value of χ^2 as a function of g_{ex}/g_{gd} for ^{133}Cs data shown in Fig. 2b and 2a from 'thin' and transmission integral analysis [Ref. 8].

4b. Quadrupole Patterns

Most of the remarks made in Sec. 4a are also applicable to the analysis of quadrupole patterns. The awareness diagrams for the Mössbauer transitions involving states with $\frac{3}{2}$ and $\frac{1}{2}$ have been discussed by various authors.[8,20] In Fig. 4 we give again the awareness diagram for this case.

For the Mössbauer transitions involving higher spins usually the quadrupole spectra are only partially resolved. The quadrupole spectra of the 37-keV ($\frac{7}{2} \rightarrow \frac{5}{2}$) resonance in ^{121}Sb is a typical example of such a situation. Errors as

large as 60% are possible in certain circumstances as shown
in Fig. 5.[13] The deduction of the asymmetry parameters η
for such cases is even more difficult. In another work[12] we
have compared the values of eQV_{zz} in some Sb compounds
obtained from the Mössbauer spectra analyzed with the 'thin'
approximation with those obtained from NQR spectroscopy.
These deviations essentially disappeared when the data were
analyzed with the transmission integral.

The case of Mössbauer transitions involving other spins
will not be dealt with here, but awareness diagrams have been
published for such measurements.[8,15]

4c. Isomer Shifts

We remark that the errors in isomer shift determinations
caused by 'thin' analysis will be small even for quite thick
samples if the spectra themselves are symmetric. Such
situations are realized, for example, in pure magnetic
hyperfine spectra. The thin analysis will, however, not be
sufficient to yield accurate isomer shift from asymmetric
hyperfine patterns, which usually involve quadrupole interaction. The effects are especially serious in the case of
$2 \rightarrow 0$ interband transitions of the rare earth, 5d, and the
actinide nuclei. The isomer shift in these cases is usually
small and many of their compounds give unresolved quadrupole
patterns. An awareness diagram for this case has been
reported by Hershkowitz et al.[15]. In Fig. 5 we have illustrated the case of the transition involving $\frac{7}{2}$-$\frac{5}{2}$ spins.

4d. Gol'danski'i-Karyagin Effects

The Gol'danski'i-Karyagin effect[21] (GKE) has now been
reported as observed in the quadrupole spectra of a large
number of nuclei. It results from the fact that $\langle x^2 \rangle$ for an
atom is generally not the same in all crystal directions;
correspondingly, the Debye-Waller resonant fraction must
vary in different crystal directions. Since the hyperfine
fields are locked to the same axes, this means that the
single-crystal line intensities are disturbed from the simple
predictions, and even after the crystals are powdered and
randomized in direction, the relative line intensities
deviate from the normalized Clebsch-Gordon coefficients W_i.
If the above effect happened to attenuate the stronger lines

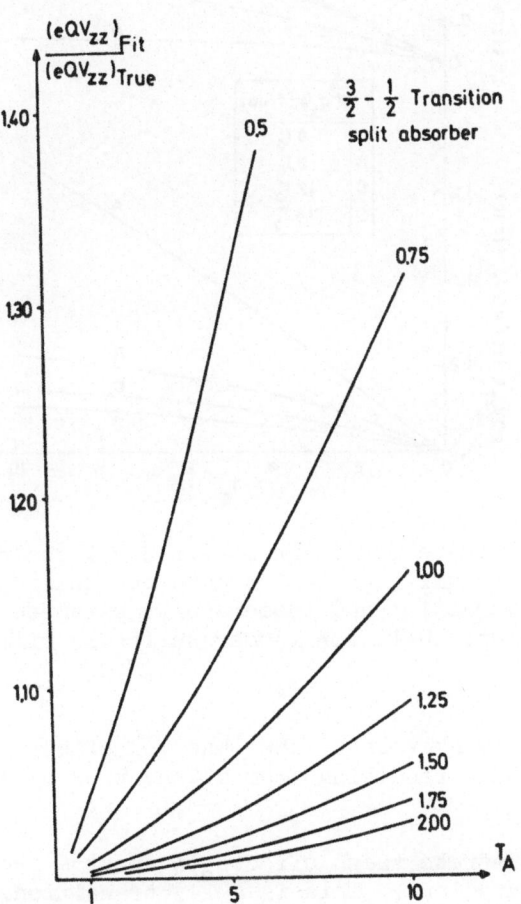

Fig. 4. Awareness diagram for the symmetric quadrupole hyperfine interaction in $\frac{3}{2} \rightarrow \frac{1}{2}$ transitions. The number on each curve gives $(eQV_{zz}/\Gamma_o)_{TRUE}$ [Ref. 8].

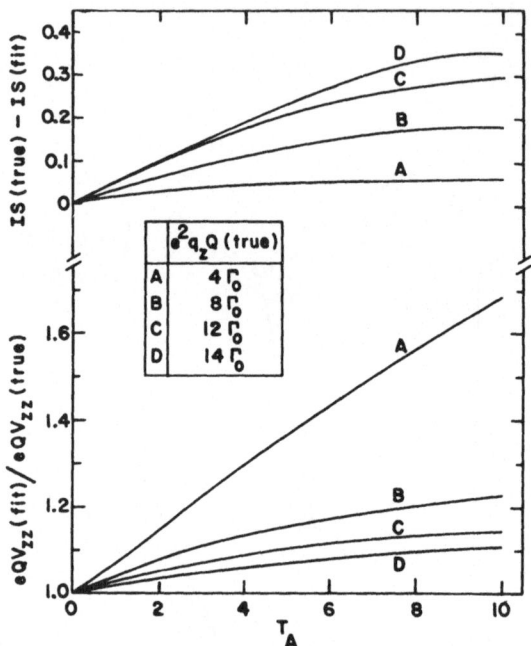

Fig. 5. Awareness diagrams for $\frac{7}{2} \to \frac{5}{2}$ quadrupole ^{121}Sb
spectrum. Errors for both the quadrupole-
interaction and isomer-shift parameters are
shown. Unit for isomer shifts is millimeter/sec.

more than the weaker ones, the observed effect would look
rather like that resulting from attenuation by a thick
absorber.

We demonstrate these effects in spectra from ^{121}Sb in an
axial field gradient. This is a $\frac{7}{2}-\frac{5}{2}$ transition. Curve a of
Fig. 6 shows the quadrupole spectrum, including GKE,[12] as
calculated in the thin approximation. Curve b of this figure
is the same without GKE, but for varying absorber thicknesses
while using the proper transmission integral. In comparison,
indeed the GKE and absorber saturation have similar effects
on the spectrum.

An experimentally measured spectrum (even with $T_a < 1$)
can be mistaken as evidence for the presence of GKE with

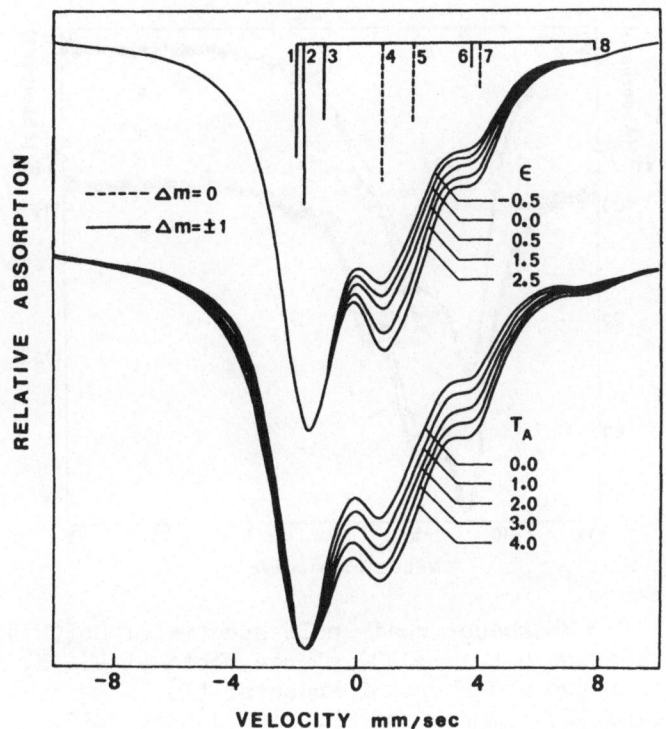

Fig. 6. a) Quadrupole spectra for ^{121}Sb with eQV_{zz} = -20 mm/sec, Γ_O = 1.05 mm/sec, η = 0, and for various values of $\varepsilon = k^2(\langle z^2 \rangle - \langle x^2 \rangle)$ calculated in the thin approximation. b) The transmission integral simulated spectra for above case with ε = 0, but for different absorber thicknesses. All spectra are normalized to the same maximum dip.

$\varepsilon = (\langle z^2 \rangle - \langle x^2 \rangle) k^2 > 0$. Here $\langle z^2 \rangle$ and $\langle x^2 \rangle$ are the mean-square displacement parallel and perpendicular to the principal symmetry axis, and k is the absolute value of the wave vector for the γ radiation.

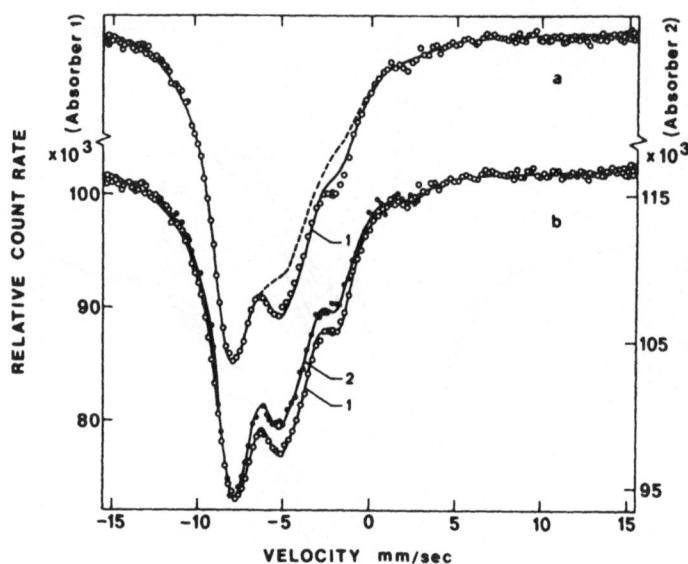

Fig. 7. Mössbauer quadrupole spectra of $Sb(CH_3)_3Br_2$
against a $Ba^{121m}SnO_3$ source, both at 4.2 K,
for two absorber thicknesses: (1) 12 mg/cm²
and (2) 7 mg/cm² of Sb. In (a) the solid
curve represents the thin approximation
including the GKE; the dashed curve, without
the GKE. In (b) the data for the two
absorbers are fitted by TI without any GKE.

The spectra of $Sb(CH_3)_3Br_2$ measured for two absorber
thicknesses against a $Ba^{121m}SnO_3$ source, both at 4.2 K is
presented in Fig. 7. In this compound the site symmetry of
Sb ions is D_{3h}. The dashed curve (a) in Fig. 7 is the best
fit obtained from a 'thin' analysis assuming $\epsilon = 0$. Inclusion
of GKE in this analysis considerably improves the fit (solid
curve (a) in Fig. 7) except in the region of line 7 (see
Fig. 6). Such an analysis with two absorber thicknesses,
however, did not yield identical values for ϵ. In Fig. 7,
curves (b), we give the fits for two absorbers using the
transmission integral (Eq. (3)) without including any GKE.
These fits are more satisfactory than those of curves (a)
on several grounds. Thus the transmission integral analysis
has removed the evidence for the presence of nonzero GKE
parameters in this compound.

The discussion given above is not unique to ^{121}Sb spectra and is likely to be realized in spectra involving different Mössbauer transitions.

Single-crystal absorbers, of course, create spectra for which the relative line intensities are different from those of powdered absorbers. If a single crystal absorber is made thick, a new effect makes its appearance. The source γ rays are actually not all of one type, but can be described as a mixture of two polarizations, say parallel and perpendicular to the principal axis of the absorber. These can have different absorption cross sections, and the transmission now must be treated as referring to the sum of the trans- missions of the two polarizations. We will not treat this problem here, and refer the interested reader to the useful and clear work of Housley et al.,[22] and of Pfannes and Gonser.[23]

4e. Interference Effects

There are two ways to produce a final state consisting of the de-excited source nucleus and an absorber atom with a hole in its k shell. One is nuclear-resonance absorption followed by k-shell internal conversion; another is the ordinary electronic-photoelectric effect. The amplitudes of these processes must be added, and when the sum is squared to find the liklihood of such processes, a cross term can result. It is called a dispersion term for it modifies the normal Lorentzian shape similarly to dispersion in resonant absorption of ordinary light. In fact, Eq. (4) is changed to

$$A(E) = \frac{\Gamma_0}{\Gamma_a} \frac{1 - 2\xi x}{1 + x^2} , \tag{8}$$

where ξ is the amplitude of the dispersion or interference term.

Such effects have been observed in E1, E2 and mixed E2/M1 transitions.[7,24,25,26] It has been observed that the dispersion effect seems to be enhanced by using thick absorbers.[26]

Erickson et al.[7] have made such analyses in connection with their study of the dispersion term in ^{197}Au. The

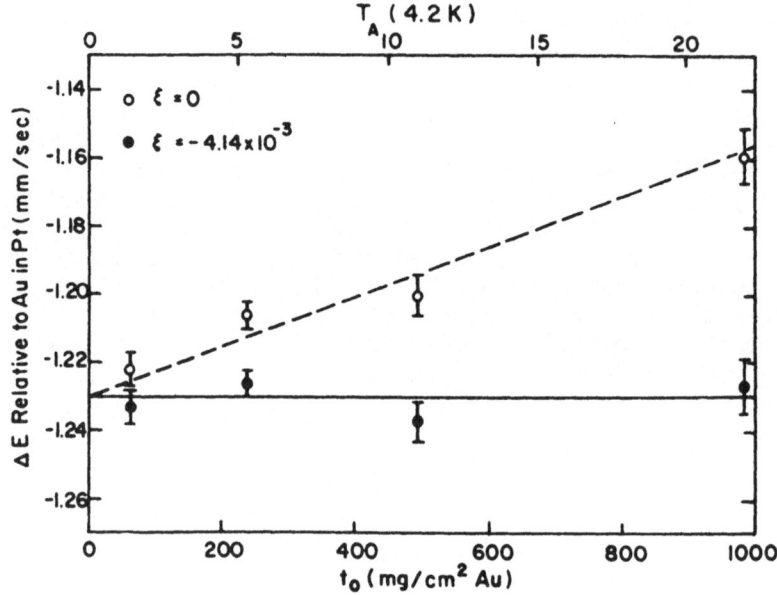

Fig. 8. Isomer shift ΔE versus Au-metal absorber
thickness. The dashed curve is 'thin' with
the interference parameter ξ = 0, while the
solid curve uses ξ = -4.14 × 10⁻³ and TI
analysis [Ref. 7].

deduced isomer shifts will be subject to error if either the
dispersion term or absorber thickness effects are neglected.
In Fig. 8, Erickson's isomer shift data for various thickness with and without the dispersion term are given. Small
isomer shifts can be readily 'created.' These effects will
be of importance for the measurement of the small isomer
shifts in the 2 → 0 transitions; here E2 transitions tend to
create relatively large values of ξ .

4f. Mössbauer Thermometer

The basis of the Mössbauer low-temperature thermometer
has been discussed by Kalvius et al.[27] The temperature is
arrived from the relative intensities of the various hyperfine lines as these are affected by the temperature-dependent
Boltzmann populations of the ground state hyperfine levels
of the resonant absorber nuclei. Among the various

Mössbauer transitions suggested for the absorber thermometer, the 21.6-keV transition in ^{151}Eu has acceptable characteristics for the temperature range from 25 to 250 mk.

A primary difficulty in the use of Mössbauer thermometers has arisen from the fact that the transmitted intensity is not linearly dependent on the absorber thickness, but in fact, requires use of the transmission integral. The absorber thickness for a particular transition in the hyperfine spectrum at low temperatures must be modified from $T_A W_1$ (see Eq. (2)) by the factor $P(m)$ where

$$P(m) = \frac{\exp(-m\,E/kT)}{\sum_{m=-Ig}^{Ig} \exp(-m\,E/kT)} \qquad (9)$$

where

$$E = g\,\mu_n H_n.$$

Here H_n is the nuclear magnetic field, g is the nuclear ground state g-factor, kT is the temperature in energy units, and m is the magnetic quantum number for the ground state of this particular nuclear transition.

In order to demonstrate the influence of absorber thickness we show first, in Fig. 9a, spectra for EuS at 4200, 40, 60, and 80 millidegrees as calculated 'thin'; second, in Fig. 9b, spectra at 50 millidegrees for T_A = 1, 4, and 6 calculated with a transmission integral. Comparison of the two indicates that the 'thin' approximation leads to over-estimates of the temperature. Further, for T_A = 10 (a reasonable thickness for an absorber split into five major absorption lines), the apparent temperature is a full 50% larger than the actual.[28]

CONCLUSIONS

We have attempted to show some of the kinds of errors that can be caused by use of the 'thin' approximation, rather than the transmission integral, in the analysis of Mössbauer data. The so-called 'awareness' plots can give one a feeling for the possible size of the error, if you have a physical case close to one already analyzed.

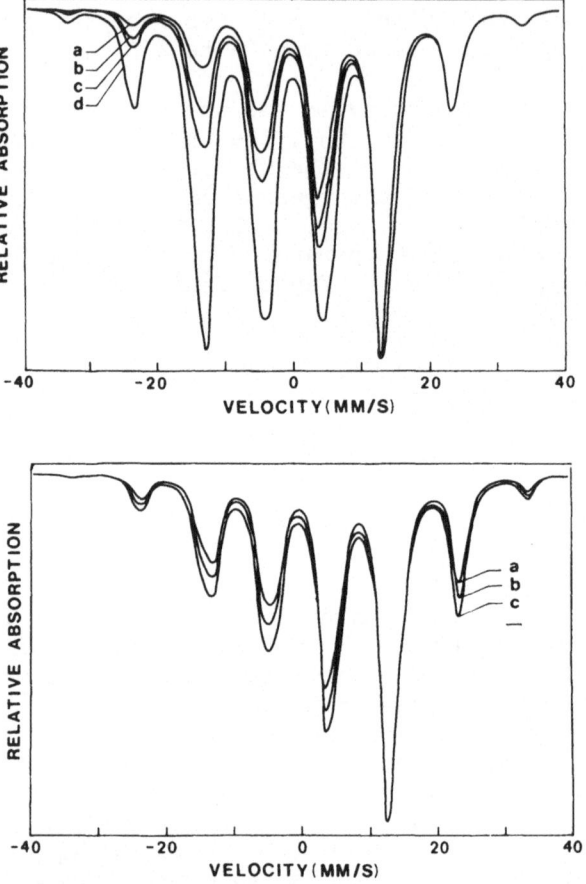

Fig. 9. (a) The hyperfine magnetic spectrum of
^{151}Eu in EuS at a)40, b) 60, c) 80 and d)
4200 mk calculated in the thin approximation.
(b) The spectrum for EuS at 50 mk and for
absorber thickness, $n_A\sigma_o f_A$, of a) 1, b) 4,
and c) 6 [Ref. 28].

There will still be a majority of cases where the
improved technique is not really needed. For example, all
well-resolved spectra can be treated as a collection of single
lines, and thickness effects can be taken into account by
older methods. Even with poorly resolved spectra, saturation
effects cannot be important if $T_A \leqslant 1$. And this condition
is enforced in many experiments by such factors as low
isotopic enrichment, large absorber dilution by other

elements, high photoelectric absorption, etc. It is, of
course, clear that a small observed effect does not guarantee
that T_A is small, as this is most often due to a small
resonant fraction in the source, or a poor signal to background ratio in the detector. Other special cases occur
where neglect of the transmission integral can be justified;
for example, the isomeric shifts in symmetrical spectra
will not be affected, even though simultaneously the magnetic
splitting $g \mu_n H$ is being ill treated. Our procedure now is
to make most of the initial fits with the transmission
integral switched off with CONVO = 0. After the fits are
becoming sensible, then CONVO = 1 switches on the slower
transmission integral calculation. A comparison of the final
values of the parameters indicates whether it is necessary
to continue with the more expensive procedure.

REFERENCES

1.   S. Margulies and J. R. Ehrman, Nucl. Instr. Methods 12,
     131 (1961); S. Margulies, P. Debrunner and H.
     Frauenfelder, Nucl. Instr. Methods 21, 217 (1963).

2.   S. L. Ruby and J. M. Hicks, Rev. Sci. Instr. B3, 27
     (1962).

3.   J. Heberle and S. Franco, Z. Naturforch. 23 a, 1439
     (1968).

4.   B. T. Cleveland, Z. Naturforsch. 27 a, 370 (1972);
     B. T. Cleveland and J. Heberle, Phys. Lett. 36A, 33
     (1971); B.T. Cleveland and J. Heberle, Phys. Lett. 40A,
     13 (1972).

5.   J. C. Love, F. E. Obenshain, and G. Czjzek, Phys. Rev.
     3, 2827 (1971).

6.   D. J. Erickson, L. D. Roberts, J. W. Burton, and J. O.
     Thomson, Phys. Rev. B 7, 2180 (1971).

7.   D. J. Erickson, J. F. Prince, L. D. Roberts, Phys. Rev.
     C 8, 1916 (1973).

8.   E. Gerdau, W. Räth and H. Winkler, Z. Physik 257, 29
     (1972).

9.   M.C.D. Ure and P.S. Flinn, Mössbauer Effect Methodology,
     Vol. 7, ed. I. J. Gruverman (Plenum Press, New York,
     1971) p. 245.

10. T. E. Cranshaw, J. Phys. E 7, 1 (1974).

11. See for example, J. Legras Methodes et techniques de l'analyse numerique (Dunod, Paris, 1971).

12. G. K. Shenoy and J. M. Friedt, Phys. Rev. Lett. 31, 419 (1973).

13. G. K. Shenoy and J. M. Friedt, Nucl. Instr. Methods (in press).

14. R. E. Meads, B.M. Place, F. W. D. Woodhams, and R. C. Clark, Nucl. Instr. Methods 98, 29 (1972).

15. N. Hershkowitz, R. D. Ruth, S. A. Wender, and A. B. Carpenter, Nucl. Instr. Methods 102, 205 (1972).

16. G. Hembree and D. C. Price, Nucl. Instr. Methods 108, 99 (1973).

17. G. K. Wertheim, V. Jaccarino, L. R. Walker, and D. N. E. Buchanan, Phys. Rev. Lett. 12, 24 (1964).

18. M. Rubinstein, G.H. Strauss, and M. B. Stearns, J. Appl. Phys. 37, 1334 (1966).

19. S. Hüfner and E. Matthias, Mössbauer Spectroscopy and its Applications (IAEA, Vienna, 1972) p. 344.

20. S. A. Wender and N. Hershkowitz, Nucl. Instr. Methods 98, 105 (1972).

21. V. L. Gol'danski'i, E. F. Makarov, and V. V. Khrapov, Phys. Letters 3, 344 (1963); S. V. Karyagin, Dokl. Akad. Nauk. SSSR 148, 1102 (1963).

22. R. M. Housley, V. Gonser and R. W. Grant, Phys. Rev. Letters 20, 1279 (1968).

23. H. D. Pfannes and U. Gonser, Appl. Phys. 1, 93 (1973).

24. C. Sauer, E. Matthias, and R. L. Mössbauer, Phys. Rev. Letters 21, 961 (1968).

25. F. E. Wagner, B. D. Dunlap, G. M. Kalvius, H. Schaller, R. Felscher, and H. Spieler, Phys. Rev. Letters 28, 530 (1972).

26. W. Henning, G. Baehre, and P. Kienle, Phys. Lett. 31B, 203 (1970).

27. G. M. Kalvius, T. E. Katila and O. V. Lounasmaa, Mössbauer Effect Methodology, Vol. 5, Ed. I. J. Gruverman (Plenum Press, N.Y., 1970) p. 237.

28. G.K. Shenoy and H. Maletta (to be published).

APPENDIX A

```
C      QUADRUPOLE PATTERN FOR 5/2-7/2 TRANSITION IN 121 SB.
C      GAUSS LEGENDRE INTEGRATION OF THE TRANSMISSION FUNCTION.....
       IMPLICIT REAL*8(A-H.O-Z)
       DIMENSION DB(800).WTT(800).E(800).TRANS(800).XD(400).YC(400)
       DIMENSION VEL(8).CG(8).U(100). WT(100).B(100)
C      NWT =NUMBER OF WEIGHTS. RQ=QUADRUPOLE MOMENT RATIO.
       DATA PI.RQ.NWT /3.14159265358979З.1.34.12/
C      U(1)....U(6) COORDINATES OF GAUSS LEGENDRE INTEGRATION.
C      WT(1)...WT(6) WEIGHTS FOR GAUSS LEGENDRE INTEGRATION.
       DATA U(1).U(2).U(3)/0.9815606.0.9041173.0.7699027/
       DATA U(4).U(5).U(6)/0.5873180.0.3678315.0.1253334/
       DATA WT(1).WT(2).WT(3)/0.0471753.0.1069393.0.1600783/
       DATA WT(4).WT(5).WT(6)/0.2031674.0.2334925.0.2491470/
       DATA CG/21..6..15..1..10..10..3..18./
C      EQ=QUADRUPOLE INTERACTION. S= ISOMER SHIFT. TA=ABSORBER THICKNESS.
       EQ=-20.0
       S=0.0
       TA=4.0
C      FSB= SOURCE RESONANCE FRACTION=SIGNAL/(SIGNAL+BACKGROUND)
       FSB=1.0
C      WDS.WDA AND WDNAT ARE SOURCE.ABSORBER AND NATURAL WIDTHS.
       WDS=1.05
       WDA=1.05
       WDNAT=1.05
       LINES=8
C      VEL(I) AND CG(I) GIVE THE LINE POSITION AND NORMALIZED INTENSITY OF
C      I TH LINE IN THE HYPERFINE SPECTRUM.
       VEL(1)=S+EQ*(RQ-1.)/4.0
       VEL(2)=S+EQ*(RQ/7.-1.)/4.0
       VEL(3)=S+EQ*(RQ/7.+0.2)/4.0
       VEL(4)=S+EQ*(RQ*(-3.0/7.0)-1.)/4.0
```

APPENDIX A (cont'd)

```
        VEL(5)=S+EQ*(RO*(-3.0/7.0)+.2)/4.0
        VEL(6)=S+EQ*(RO*(-3.0/7.0)+0.8)/4.0
        VEL(7)=S+EQ*(RO*(-5./7.)+0.2)/4.0
        VEL(8)=S+EQ*(RO*(-5./7.)+0.8)/4.0
        DO 2 I=1,8
      2 CG(I)=CG(I)*TA*WDNAT/(WDA*84.)
C     VEXPT IS THE EXPERIMENTAL VELOCITY.
        VEXPT=20.0
C     NP = NUMBER OF DATA POINTS.
        NP = 199
        HNP=(NP/2.0)+0.5
C     XD(I)= VELOCITY OF I TH CHANNEL.
        DO 5 I = 1,NP
      5 XD(I)=VEXPT*(I-HNP)/HNP
        WDEXPT=WDS+WDA
C     ASSIGNING WEIGHTS TO THE NEGATIVE COORDINATES.
        M=NWT/2
        DO 10 I=1,M
        WT(NWT-I+1)=WT(I)
     10 U(NWT-I+1)=-U(I)
        WDSSO= (WDS/2.)**2
        WDASO= (WDA/2.)**2
C     ZONEND= WIDTH OF THE INNER ZONE.
        ZONEND= VEXPT+5.*WDEXPT
C     NZONE = NUMBER OF SUBZONES IN THE INNER REGION.
C     NADSUB=NUMBER OF SUBZONES TO BE ADDED OR TO BE SUBTRACTED FROM NZONE
        NZONE=ZONEND/WDEXPT
C     NADSUB= PARAMETER TO INCREASE OR DECREASE THE NUMBER OF SUBZONES.
        NADSUB=0
```

APPENDIX A (cont'd)

```
      NZONE=NZONE+NADSUB
C   NUBER OF SUBZONES IN THE WING=NWING
      NWING=5
C   NZONET= TOTAL NUMBER OF SUB ZONES IN ONE-HALF OF THE SPECTRUM.
      NZONET=NZONE+NWING
C   DEFINING THE BOUNDARIES OF THE SUBZONES IN THE WINGS.
      BMIN=0.0
      DO 21 K=1,NWING
21    BMIN=BMIN+10.**K
C   B(1) = VALUE OF THE NEGATIVE INFINITY FOR THE INTEGRATION.
      B(1)=-ZONEND-BMIN*NDEXPT
      DO 15 I=1,NWING
      K=NWING+1-I
15    B(I+1)=B(I)+NDEXPT*10.**K
C   DEFINING THE BOUNDARIES OF THE INNER SUB ZONES.
C   WDFACT= WIDTH OF INNER SUB ZONE.
      WDFACT=ZONEND/NZONE
      DO 16 I=1,NZONE
      K=NWING+I
16    B(K+1)=B(K)+WDFACT
C   DEFINING SUB ZONE BOUNDARIES IN THE REMAINING HALF OF THE SPECTRUM.
      DO 17 I=1,NZONET
      K=NZONET+1-I
      L=NZONET+1+I
17    B(L)=-B(K)
C   CALCULATION OF GAUSS-LEGENDRE COORDINATES AND TRANS.
C   NZONE2 TOTAL NUMBER OF SUBZONES OVER ENTIRE SPECTRUM.
      NZONE2=NZONET*2
      NSTEP=0
      DO 30 L=1,NZONE2
```

APPENDIX A (cont 'd)

```
      DO 30 K=1,NWT
      NSTEP=NSTEP+1
      DB(NSTEP)=(B(L+1)-B(L))/2.
      E(NSTEP)=U(K)*DB(NSTEP)+(B(L+1)+B(L))/2.
      WTT(NSTEP)=WT(K)
      ABS=0.0
      DO 25 K1=1,LINES
      BA=(E(NSTEP)-VEL(K1))**2+WDASQ
   25 ABS=ABS+CG(K1)*WDASQ/BA
   30 TRANS(NSTEP)=DEXP(-ABS)
C     NSTEPS= NUMBER OF STEPS OVER WHICH TRANS HAS BEEN COMPUTED.
      NSTEPS=NWT*NZONE2
C     YØ=COUNT RATE AT INFINITE VELOCITY.
      YØ=1.ØD6
C     CONVOLUTION LOOP...
C     YC(I)= CALCULATED COUNTS IN CHANNEL I.
      DO 40 I=1,NP
      YC(I)=Ø.Ø
      DO 35 L=1,NSTEPS
      ESP=(E(L)-XD(I))**2
      FLUX=(WDS/(2.*PI))*(1./(ESP+WDSSQ))
      R=FLUX*TRANS(L)
   35 YC(I)=YC(I)+R*WTT(L)*DB(L)
   40 YC(I)= YØ* ((1.-FSB)+FSB*YC(I))
      PRINT 5Ø. (I,XD(I),YC(I), I=1,NP)
   50 FORMAT(1HØ,16.2X,2G12.6,16.2X,2G12.6)
      STOP
      END
```

APPENDIX B

```
C     CRANSHAW TRANSMISSION INTEGRAL FOR SINGLE SITE ANTIMONY.   QUAD-
C     RUPOLE ONLY: NP ODD OR EVEN.   CONVO FLAG MUST BE ZERO OR ONE.  Q
C     AND QP ARE CENTER POINTS IN TERMS LIKE V=DV*(I-Q).  N IS THE DUMMY
C     ENERGY VARIABLE OF INTEGRATION; M REFERS TO DIFFERENT VELOCITIES
C
      DIMENSION XD(400),YD(400),YC(400)
      DIMENSION F(402),X(20),VEL(4,8),CG(8)
      DIMENSION SUM(400),WING(400),SUM(400)
      DIMENSION SUM(400),FLUXZP(800),WING(400),SUM1(400)
      DATA CG/21.,6.,15.,1.,10.,3.,18./
C
      CONVO = 1
      NP = 199
      FNP= NP
      NPHALF=NP/2
      Q = (NP/2.0) + 0.5
      DQ = (NP/2.0) - NP/2
      NP2 = 2*NP + 2*DQ
      QP = NP2/2.0 + 0.5
      PI=DARCOS(-1.0D0)
C
      DV = 0.2
      WDNAT = 1.05
      WDS = 1.05
      FSB = 1.0
      RQ = 1.34
      YQ = 1000000.0
      TA = 4.0
      NDA = 1.05
      S = 0.0
      EQ = -20.0
      VEL(1) = S + (EQ/4.0)*(RQ - 1.0)
```

APPENDIX B (cont 'd)

```
      VEL(2) = S + (EQ/4.0)*(RQ/7.0 - 1.0)
      VEL(3) = S + (EQ/4.0)*(RQ/7.0 + 0.2)
      VEL(4) = S + (EQ/4.0)*(RQ(-3.0/7.0) - 1.0)
      VEL(5) = S + (EQ/4.0)*(RQ(-3.0/7.0)+ 0.2)
      VEL(6) = S + (EQ/4.0)*(RQ(-3.0/7.0)+ 0.8)
      VEL(7) = S + (EQ/4.0)*(RQ(-5.0/7.0) +0.2)
      VEL(8) = S + (EQ/4.0)*(RQ(-5.0/7.0) +0.8)
C
C   TRANSMISSION: FOR SINGLE LINE A(I) = TA*(WDNAT/WDA)/(1 +
C   ((V-VP)/WDA/2))**2).  FOR J LINES.  JUST SUM USING CG FOR THE UN-
C   NORMALIZED HEIGHTS.  WHEN CONVO =0. WIDTH BECOMES WDA+WDS..  IF SHAPE
C   IS OTHER THAN LORENTZIAN. NORMALIZE SO INTEGRAL OF A IS (WDNAT*PI/I).
      NPSTOP = NP2*CONVO + NP*(1-CONVO)
      GAMMA = (1-CONVO)*(WDS+WDA) + CONVO*WDA
      DO 24 I = 1. NPSTOP
      A(I) = 0.0
      FI = I
      V1(I) = DV*(FI - ((1-CONVO)*0 + CONVO*QP))
      DO 22 K= 1.8
      WT = (CG(K)/84.0)*(WDNAT/GAMMA)
  22  A(I) = A(I) + WT/(1.0 + ((V1(I) -VEL(1.K))/(GAMMA/2.0))**2)
  24  TRANS(I) = DEXP(-TA*A(I))
      IF (CONVO .EQ. 1) GOTO 28
      DO 27 I=1.NP
      YC(I) = X(1)*( 1.0 - FSB + FSB*TRANS(I))
  27  F(I) = (YD(I) - YC(I))/DSQRT(YD(I))
      RETURN
C
```

APPENDIX B (cont'd)

```
C  CALCULATION OF FLUXZ
28 FAC = 2.0*DV/WDS
   DO 30 L=1,NP2
   FL = L
30 FLUXZ(L) = (2.0/(PI*WDS))/(1.0+ (FAC*(FL-FNP))**2)
C
C  CONVOLUTION CALCULATION
   DO 50 M = 1,NP
   SUM1(M)=0.0
   NSTART=M-NPHALF+1
   IF(NSTART.LT.1)NSTART=1
   NSTOP=M+NP2-NPHALF
   IF(NSTOP.GT.NP2) NSTOP=NP2
   DELM = DATAN(FAC*(NSTART-M-NPHALF-0.5-2.0*DQ))
   DELP = DATAN(FAC*(NSTOP-M-NPHALF +0.5-2.0*DQ))
   WING(M) = (1.0/PI)*(DELM-DELP) + 1.0
   DO 40 N=NSTART,NSTOP
40 SUM1(M)=SUM1(M) + TRANS(N)*FLUXZ(N+NPHALF-M)
   SUM(M) = DV*SUM1(M) + WING(M)
50 YC(J) = Y0*(1.0 - FSB*(1.0-SUM(J)))
   F(J) = (YD(J) -YC(J))/DSQRT(YD(J))
   RETURN
C
15 PRINT 940, (I,XD(I),YC(I),I=1,NP)
940 FORMAT(1H0,I6,2X,2G12.6,I6,2G12.6)
   RETURN
   END
```

SELECTIVE EXCITATION DOUBLE MÖSSBAUER SPECTROSCOPY*

Bohdan Balko and Gilbert R. Hoy

Physics Department, Boston University

Boston, Massachusetts 02215

INTRODUCTION

Although the selective excitation double Mössbauer effect (SEDM) has been known for some time,[1,2,3] it has only recently been applied to the study of solid state problems[4]. The reason for the delay is that the necessary theory and computational techniques had to be developed to handle the case of thick specimens which are required for an adequate counting rate.

The reason that SEDM experiments are so interesting is that, in contrast with Mössbauer transmission and ordinary scattering experiments which measure the absorption cross section as a function of the incoming energy, these new experiments measure the differential scattering cross section as a function of both the incoming and outgoing energies. In this way not only the absorption

---

*This work was supported by the National Science Foundation under Grant No. GH-39137.

energy is determined, but also the spectral energy distribution of the emitted radiation. Thus in principle any energetic interactions of the excited nucleus with its environment during its lifetime can be detected.

The characteristic feature of SEDM experiments is that four recoilless events constitute a measurement, and therefore two Doppler modulators have to be used to test for their occurrence. A block diagram of the SEDM apparatus is shown in Figure 1. The components comprising the Doppler modulators are of conventional design and utilize the electromechanical driving technique. Drive 1 is a constant velocity drive (CVD) which is set to Doppler shift the incoming gamma ray to the desired energy. The incoming photons are scattered by the "scatterer", which is made of the material under study. The re-emitted radiation is energy analyzed by the "analyzer" which is driven in a constant acceleration mode (CAD). The resulting energy spectral distribution is accumulated in a multichannel analyzer using standard procedures.

The experiments to be described utilized the Mössbauer effect in ^{57}Fe. The source, purchased from New England Nuclear, was 45mC of ^{57}Co in a copper foil. The scatterers were made from commercially available 90% enriched and natural α-Fe_2O_3 and Fe metal powders. The analyzer was a single line $Na_4Fe(CN)_6 \cdot 10H_2O$ absorber having a thickness parameter $\beta = 14$ ($\beta = n''\sigma_o f$ where n'' = number of ^{57}Fe atoms/cm^2, σ_o = resonance cross section, and f = the recoilless fraction). Because of the need to prevent unwanted radiation from reaching the detector (a proportional counter), special copper clad lead shielding was used.

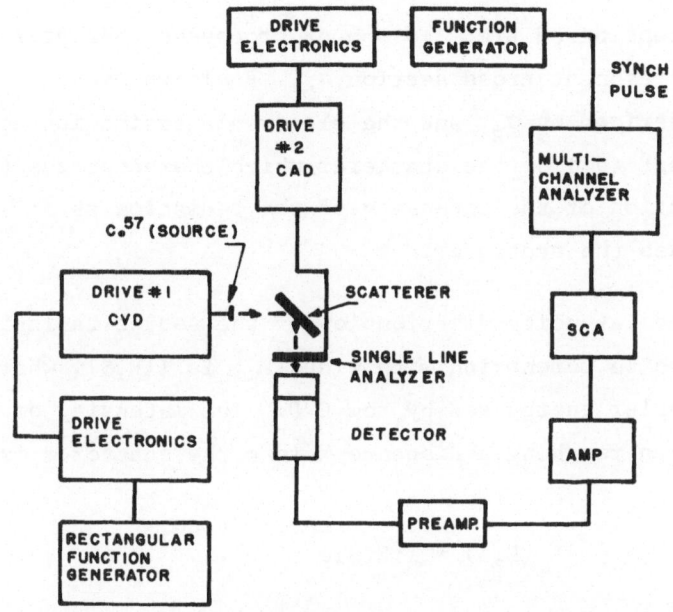

Fig. 1.   Schematic block diagram of the SEDM apparatus.

In analyzing SEDM experiments it is important to know
the energy distribution of the source radiation.   This distribution, which includes thickness effects in the source,
finite source area, collimation, and mechanical distortions,
was measured by conducting constant velocity transmission
experiments using Drive 1 and a thin absorber.   The
measured effective linewidth of our source was $\Gamma_b = 2.3\Gamma$
where $\Gamma = 0.097$ mm/sec.

SEDM RESULTS ON A TIME INDEPENDENT ENVIRONMENT

In this section we calculate the energy spectral distribution of radiation that reaches the detector after
being scattered by a "split", "thick" scatterer and transmitted through the analyzer.   The major effects that need

to be considered are; Mössbauer processes characterized
by the resonant cross section σ_o, Rayleigh events
characterized by σ_R, and the electronic extinction coefficient (μ_e) of the scatterer which characterizes the
attenuation of the intensity of the radiation as it
traverses the scatterer.

The intensity distribution of the source radiation,
taken to be Lorentzian with width Γ_b, is $I(E,S)$, where S is
the Doppler energy set by the CVD. The intensity of the
radiation reaching a distance x into the scatterer is

$$I_x(E,S) = I(E,S)e^{-\mu_T(E)\csc\alpha_1 x} \tag{1}$$

where α_1 is the angle of the incoming beam relative to the
scatterer and $\mu_T(E)$ is the scatterer's total absorption coefficient. Thus $\mu_T(E)$ can be expressed as,

$$\mu_T(E) = \mu_M(E) + \mu_R + \mu_e \tag{2}$$

where $\mu_M(E)$, μ_R, and μ_e are the nuclear resonant, Rayleigh
and electronic absorption contributions, respectively.

The resonant nuclear absorption can take place at
energies corresponding to the allowed nuclear transitions
E_{ij}. If the nuclear energy levels are pure m states,

$$\mu_M(E) = \sum_{i,j} \frac{nf\sigma_o(\frac{\Gamma}{2})^2}{(E-E_{ij})^2 + (\frac{\Gamma}{2})^2} W_{ij}(\theta_1\phi_1) \tag{3}$$

where f is the scatterer's recoilless fraction, n is the

number of resonant nuclei per unit volume, Γ is the natural line width, and σ_o is the maximum resonant absorption cross section. The energy levels are defined in Fig. 2. The functions $W_{ij}(\theta_1,\phi_1)$ are

$$W_{ij} = |C(I_e m_j | LMI_g m_i)|^2 |\vec{X}_L^M(\theta_1,\phi_1)|^2 \tag{4}$$

Here the C's are the usual Clebsch-Gordan coefficients and the $\vec{X}_L^M$'s are vector spherical harmonics.

The energy distribution per unit length per unit solid angle of the radiation scattered at x by resonant nuclear processes is

$$\frac{d}{dx}I_M(E',S) = \int dE I_x(E,S) nf \frac{d^2\sigma_M}{dEdE'} \tag{5}$$

Following Heitler[5], and Boyle and Hall[6] we can write the differential scattering cross section as

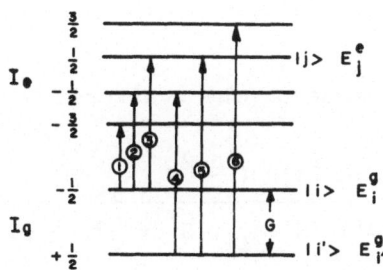

Fig. 2. Energy level diagram for a "split" [57]Fe iron scatterer. The allowed transitions are labeled 1,2,3,4, 5 and 6.

$$\frac{d^2\sigma_M}{dEdE'} = \sum_{i,j,i'} \frac{f\sigma_o W_{ij}(\theta_1\phi_1)W_{ji'}(\theta_2\phi_2)}{(1+\alpha')[(E-E'+\Delta_{ii'})^2+(\frac{\gamma}{2})^2][(E'-E_{ij}-\Delta_{ii'})^2+(\frac{\Gamma}{2})^2]} \tag{6}$$

where α' is the internal conversion coefficient, γ is the
line width due to the source intensity and hence is very
small, and we have allowed for absorption between the
nuclear levels $E_i^g \to E_j^e$ to result in the decays $E_j^e \to E_i^g$
and $E_j^e \to E_{i'}^g$. $E_{ij} = E_j^e - E_i^g$, and $\Delta_{ii'} = E_i^g - E_{i'}^g = G$
(see Fig. 2). In the above sum we must include all transi-
tions consistent with the selection rules.

Substituting Eq. 6 and Eq. 1 into Eq. 5 and integrat-
ing over E gives   (7)

$$\frac{dI_M}{dx}(E',S) = \sum_{i,j,i'} \frac{nf^2\sigma_o I_o(\frac{\Gamma_b}{2})^2 e^{-\mu_T(E'-\Delta_{ii'})\csc\alpha_1 x} W_{ij}W_{ji'}}{(1+\alpha')[(E'-\Delta_{ii'}-S)^2+(\frac{\Gamma_b}{2})^2][(E'-E_{ij}-\Delta_{ii'})^2+(\frac{\Gamma}{2})^2}$$

The energy distribution of the nuclear resonant scattered
radiation that gets to the analyzer is

$$I_M(E',S) = \int_o^T \frac{d}{dx}I_M(E'S)e^{-\mu_T(E')\csc\alpha_2 x} dx$$

where T is the thickness of the scatterer. The result is

$$I_M(E',S) = \sum_{i,j,i'} \frac{nf^2\sigma_o I_o(\frac{\Gamma_b}{2})^2 W_{ij}W_{ji'}F}{(1+\alpha')[(E'-\Delta_{ii'}-S)^2+(\frac{\Gamma_b}{2})^2][(E'-E_{ij}-\Delta_{ii'})^2+(\frac{\Gamma}{2})^2} \tag{8}$$

where $F = \dfrac{1-e^{-T(\mu_T(E'-\Delta_{ii'})\csc\alpha_1+\mu_T(E')\csc\alpha_2)}}{\mu_T(E'-\Delta_{ii'})\csc\alpha_1 + \mu_T(E')\csc\alpha_2}$

Eq. 8 is essentially the same as found in Debrunner and Morrison[7] but generalized to a "split" scatterer. The angular distribution of the resonantly scattered radiation is represented by $W_{ij}(\theta_1,\phi_1)W_{ji'}(\theta_2,\phi_2)$. If the scatterer is a powder, this factor must be averaged subject to the constraint that the scattering angle α is constant. These values have been calculated[8].

The coherent non-resonant contribution to the scattered intensity is mostly Rayleigh, because at these energies the Rayleigh contribution is $\sim 10^4$ times the Thomson contribution[9]. In general the energy distribution of the scattered radiation contains an interference term from the nuclear resonant and Rayleigh scattering processes. However, at a scattering angle $\alpha = 90°$ this interference term is zero[10]. We restrict our calculations and experiments to this case. Under these conditions the terms can be calculated separately and added. The energy dependence of the Rayleigh cross-section is relatively weak and so

(9)

$$I_R(E',S) = \frac{I_o(\frac{\Gamma_b}{2})^2 n' f_R \sigma_R W_R(\alpha)}{[(E'-S)^2 + (\frac{\Gamma_b}{2})^2]} \left[\frac{1-e^{-\mu_T(E')(\csc\alpha_1+\csc\alpha_2)T}}{\mu_T(E')(\csc\alpha_1 + \csc\alpha_2)}\right]$$

where n' is the number of atoms/vol, f_R is the recoilless fraction for Rayleigh scattering ($f_R = f^2$ for $\alpha = 90°$)[11], σ_R is the Rayleigh cross-section evaluated at 14 kev (for our purposes), and[9]

$$W_R(\alpha) = \frac{1 + \cos^2\alpha}{\sin^3\frac{\alpha}{2}} \, .$$

Collecting energy independent factors,

$$I_R(E',S) = \frac{n'A_R[1-e^{-\mu_T(E')(\csc\alpha_1 + \csc\alpha_2)T}]}{[(E'-S)^2 + (\frac{\Gamma_b}{2})^2]\mu_T(E')(\csc\alpha_1 + \csc\alpha_2)} \tag{10}$$

The total energy dependent intensity distribution of the scattered radiation, which is subject to analysis by the analyzer in the usual transmission geometry, is

$$I(E',S) = I_M(E',S) + I_R(E',S) \tag{11}$$

Once $I(E',S)$ is determined the effect of the analyzer can be calculated by

$$I(S,S') = \int_o^\infty I(E',S)e^{-\mu_a(E',S')T_a}dE' \tag{12}$$

where S' is the Doppler energy of the "single line" analyzer and

$$\mu_a = \frac{n_a f_a \sigma_o (\frac{\Gamma}{2})^2}{[(E'-S'-E_o)^2 + (\frac{\Gamma}{2})^2]} = \frac{\beta_a (\frac{\Gamma}{2})^2}{[(E'-S'-E_o)^2+(\frac{\Gamma}{2})^2]T_a}$$

Computer programs have been written to solve the transmission integral (Ruby and Hicks[12]) given by Eq. 12.

All our experiments were done using [57]Fe, and the calculations were performed assuming that the nuclear levels are pure m states. As noted above $\alpha = 90°$ and $\alpha_1 = \alpha_2 = 45°$. The E_{ij}'s and G can be determined by ordinary transmission experiments. The linear absorption coefficient, $\mu = \mu_R + \mu_e$, was determined experimentally for both iron and α-Fe_2O_3 samples by measuring the attenuation of the gamma ray beam on passing through various sample thicknesses. The values obtained were; $\mu = 205$ cm^{-1} for iron and $\mu = 54$ cm^{-1} for α-Fe$_2$O$_3$. The relative strength of Rayleigh to Mössbauer processes can be calculated[9]. The Rayleigh contribution can also be measured experimentally, as will be discussed below.

Before presenting the results of these calculations, it is perhaps worthwhile to consider effects that have not been included. For example, in a real experiment the scattering angle can not be restricted to 90° because of solid angle effects. Therefore cancellation of the Mössbauer and Rayleigh interference term is not complete. We have made estimates of this contribution which for our geometry is less than 1%. We have also made numerical estimates of multiple scattering processes. This too is less than 1%. We have also neglected so called "coherent thickness" effects[13] where one allows for the possibility of the photon to interact resonantly with more than one scattering center. These effects are quite small and have only been seen using single crystals at Bragg angles.

Fig. 3 shows calculated results for a hypothetical scatterer in which the hyperfine field is zero i.e. the absorption spectrum contains a single line, and the only process is the Mössbauer resonant effect i.e. $\mu_e = \sigma_R = 0$. In this calculation the CVD is set to excite the system three natural line widths off resonance. There are several observations to be made. First, as the thickness of the scatterer increases i.e. as β increases, the resulting emission line shape is peaked at the excitation energy. On

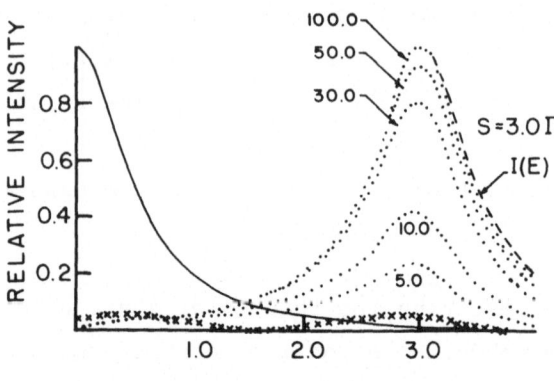

ENERGY (IN UNITS OF Γ)

Fig. 3. Calculated emission line shapes for ideal resonant scatterers. The numbers labeling the curves indicate the thickness (β) of the hypothetical scatterers. The setting of the CVD, i.e. the Doppler shift, is 3Γ.

the other hand, as β decreases we approach the "single nucleus" limit as calculated classically by Moon[14] and emphasized by Boyle and Hall[6]. In this limit the emission line shape has two peaks, one at the excitation energy and the other at the resonance energy. However, both peaks are extremely small compared to the results for thicker scatterers.

The calculated SEDM line shape for a real iron powder is shown in Fig. 4 when the sixth line is excited on resonance and at higher energies. In these calculations we include Rayleigh and incoherent thickness effects. The first point to notice is that in all cases the peak of the SEDM line comes at the energy of the incoming beam as set by the CVD. These calculations were made for two different types of samples: (1) a 90% enriched iron powder scatterer and (2) a natural iron powder scatterer both having a β = 286.

Notice in Fig. 4 a difference between the calculated results
is seen for the two different samples only for excitation
10Γ off resonance. Whereas the line shapes for (1) and (2)
are very similar, the main contributions to the two lines
are different. For case (1), i.e. the enriched iron powder,
95% of the peak is due to Mössbauer scattering and 5% to
Rayleigh scattering. For case (2), i.e. the natural iron
powder, only 25% is due to Mössbauer scattering, and the
rest is due to Rayleigh scattering. Notice also that, as a
result, case (1) is slightly less symmetrical than case (2).
The difference between the relative contributions of Mössbauer and Rayleigh processes is more clearly exhibited in
Fig. 5. In this figure we assume that the fifth line in an
iron powder scatterer is excited. In this case only the
peaks on the right contain a Rayleigh contribution. Cases 1
and 2 correspond to the same type of scatterer as assumed
for Fig. 4. Now, however, the SEDM spectrum consists of two
peaks due to the magnetic dipole selection rule. The ratio
of the two peak intensities for the enriched sample changes
only slightly in cases a, b, and c. However, for the natural sample the change in the ratio is much larger, reflecting the increased relative importance of the Rayleigh contribution. It is clear that an experiment such as this can
determine σ_R. We see that Mössbauer processes dominate
Rayleigh effects for enriched scatterers excited off resonance by less than 10Γ. However, for natural scatterers excited 10Γ off resonance, the SEDM peak is almost all due to
Rayleigh scattering. This can be clearly seen by looking
at the peak on the right side of Fig. 5 for case 2. We have
indicated by 2' the pure Mössbauer contribution to case 2.
Notice, that 2' gives a hint of a peak at the resonance

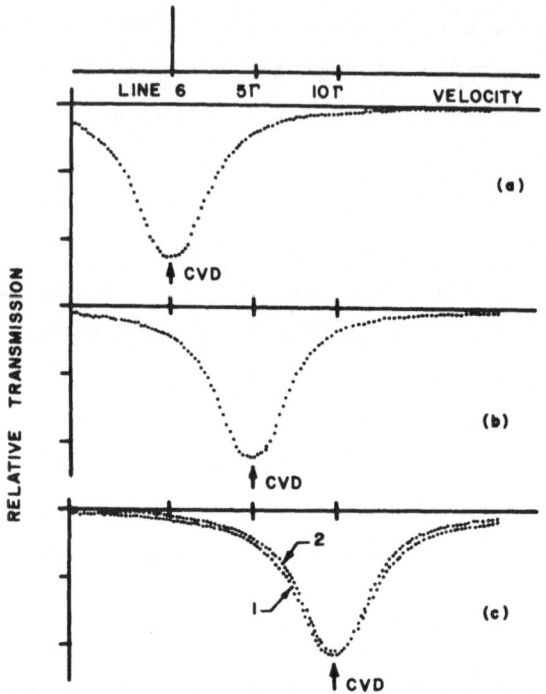

Fig. 4. Calculated SEDM results for an iron powder scatterer
when the sixth line is excited (a) on resonance, (b) 5Γ off
resonance, and (c) 10Γ off resonance. In (c) curve 1 is
for a 90% enriched sample and curve 2 is for a natural sample.

position of line 5, but these effects are completely obliterated by the large Rayleigh contribution. We have experimental results for a natural iron powder sample excited on
resonance, and an enriched iron powder sample excited off
resonance, which substantiate these calculations.

In order to check our calculations further, we performed SEDM experiments on enriched iron powder at room
temperature (298°K) and on enriched α-Fe$_2$O$_3$ powder at liquid
nitrogen temperature (77°K). The CVD was set to excite the
sixth line on resonance and two values off resonance for

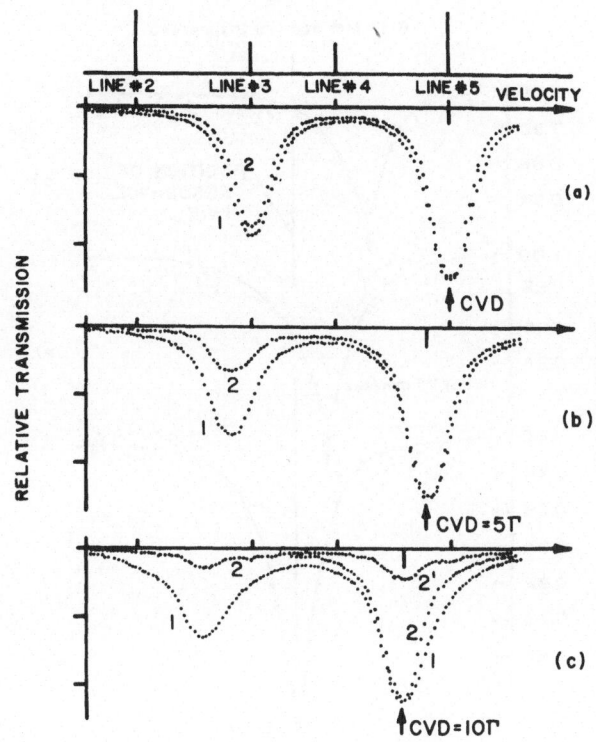

Fig. 5. Calculated SEDM results for an iron powder scatterer when the fifth line is excited (a) on resonance, (b) 5Γ off resonance, and (c) 10Γ off resonance. The curves labeled 1 correspond to a 90% enriched sample and curves 2 to a natural sample. Curve 2' in (c) shows the nuclear resonant contribution to the peak.

each of these samples. In Fig. 6 we show our results for α-Fe$_2$O$_3$ at liquid nitrogen temperature. The dots are the experimental results. The solid curves give our calculated results. The only free parameter in these calculations is the per cent effect for one spectrum. Notice that the theory correctly predicts the peak position and the total line shape.

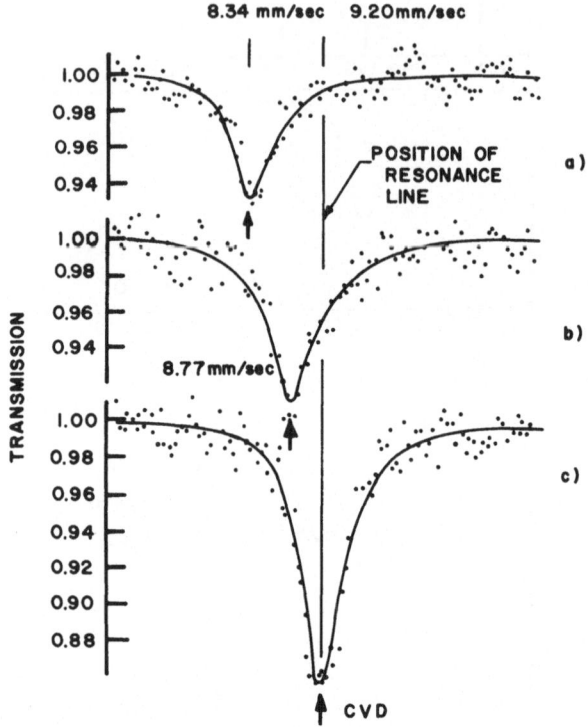

Fig. 6.   Experimental SEDM spectra for an enriched α-Fe$_2$O$_3$ scatterer at 77°K.   In (a) line 6 is excited 8.85Γ off resonance, (b) line 6 is excited 4.43Γ off resonance, and (c) line 6 is excited on resonance.

SEDM RESULTS IN A FLUCTUATING ENVIRONMENT

If the effective fields at the ^{57}Fe nuclei in the scatterer are time dependent, the expected SEDM result is different from that discussed in the previous section.   To obtain the SEDM line shape with relaxation occurring in the scatterer, the effect of the time dependent fields has to be included in the expression for the scattering line shape given in Eq. 8.   This can perhaps be done following Blume's

procedure[15]. In such an approach the effective hyperfine
fields are written as explicit functions of time by introducing a stochastic model. The central part of the calculation
then involves the determination of the correlation function
as described by Anderson[16] for the case of transmission
geometry. To find the scattering line shape the appropriate
stochastic average including these time dependent fields,
has to be obtained. In general, the SEDM problem is a more
difficult problem than the calculation for the transmission
case, and has not yet been solved. We will use a simpler
approach which is only valid for excitation on resonance.
In this method we made the necessary modifications of the
well known emission line shape formula needed to describe
the SEDM case. Following Abragam[17] we write for the
emission line shape as a function of energy ω

$$I(\omega) = \mathrm{Re}[\vec{W} \cdot \underset{\sim}{A}^{-1} \cdot \vec{I}] \tag{13}$$

where; $\vec{W}$ is the probability vector for the ionic states
which produce effective fields at the ^{57}Fe nuclei and is
time independent for an equilibrium situation, $\vec{I}$ is a unit
vector, $\underset{\sim}{A} = i(\omega - \underset{\sim}{\omega}\underset{\sim}{1}) + \underset{\sim}{\pi}$, $\underset{\sim}{1}$ is the unit matrix, $\underset{\sim}{\pi}$ is the
probability matrix for transitions between the ionic states,
and $\underset{\sim}{\omega}$ is the diagonal energy matrix.

Consider the ordinary transmission situation and suppose also a two level ionic system with both states equally
probable. Then Abragam[17] shows that

$$\underset{\sim}{\pi} = \begin{pmatrix} -\Omega & \Omega \\ \Omega & -\Omega \end{pmatrix} \tag{14}$$

where Ω is the relaxation rate and

$$
\underset{\sim}{A}^{-1} = \frac{1}{\det \underset{\sim}{A}}
\begin{pmatrix}
-i(\omega + \delta) - \Omega & \Omega \\
-\Omega & 1(-\omega + \delta) - \Omega
\end{pmatrix}
\tag{15}
$$

where $\pm\delta$ are the eigenvalues of $\underset{\sim}{\omega}$ i.e. 2δ equals the nuclear energy difference corresponding to the two possible ionic states.

The line shape for this case can be found by using Eqs. 13, 14, & 15 and noting that $W = (\frac{1}{2}, \frac{1}{2})$. The result is

$$
I(\omega) = \frac{-2\Omega \, \delta^2}{(\delta^2 - \omega^2) + 4\omega^2\Omega^2}
\tag{16}
$$

We now want to calculate a relaxation line shape expression appropriate for SEDM experiments. Consider an experiment where the CVD is set to excite one of the lines in the scatterer (we are assuming here a situation where even with relaxation they can still be resolved), say the one located at $\omega = +\delta$, corresponding to the ionic spin value $S_z = -1/2$. If there were no relaxation or $\Omega \sim 0$ we would expect the reemitted radiation to have a peak at $\omega = +\delta$ only. For $\Omega > 0$ a peak should also appear at $\omega = -\delta$.

In the analysis of this situation, it is still appropriate to consider the ionic spin flipping process to be Markovian. It is not stationary, however, since the ionic spin probabilities change with time.

If at time t=0, the nucleus is excited and sees a field corresponding to S_z = +1/2, what is the probability of its seeing the field due to S_z = -1/2 at a later time? We know from Markov process calculations[18] that

$$\vec{W}(t) = \begin{pmatrix} P_I(t) \\ \\ P_{II}(t) \end{pmatrix} = \begin{pmatrix} \frac{1}{2}(1 + e^{-2\Omega t}) \\ \\ \frac{1}{2}(1 - e^{-2\Omega t}) \end{pmatrix} \tag{17}$$

when only one state is initially populated but the states are equally likely at equilibrium. Thus P_{II} increases, while P_I decreases with time until each equals 1/2.

To obtain the average probabilities we need to average Eq. 17 over the nuclear lifetime. The result is

$$W = \begin{bmatrix} \dfrac{\Omega + \Gamma}{(2\Omega + \Gamma)} \\ \\ \dfrac{\Omega}{(2\Omega + \Gamma)} \end{bmatrix} = \begin{bmatrix} W_1 \\ \\ W_2 \end{bmatrix} \tag{18}$$

For our purposes we must generalize Eq. 18 to include the case when the equilibrium populations of the two ionic states are not equal i.e. (P_I/P_{II}) = k. This gives

$$\overline{W} = \begin{bmatrix} \dfrac{k\Omega + \Gamma}{\Gamma + (1+k)\Omega} \\ \\ \dfrac{\Omega}{\Gamma + (1+k)\Omega} \end{bmatrix} \tag{19}$$

In addition to the above, we need to modify $\underset{\sim}{A}^{-1}$
(Eq. 15) to include the natural line width Γ as discussed
by Abragam[17].

The result for the selectively excited emission line
shape including relaxation between two ionic levels is

$$I(\omega) = \frac{(\Delta^- - a\Delta^+)[\Gamma\omega + \Omega(\Delta^+ - a\Delta^-) - (1+a)\gamma(\Delta^+\Delta^- + \frac{\Gamma\gamma}{2})]}{(\Delta^+\Delta^- + \frac{\Gamma\gamma}{2})^2 + [\Gamma\omega + \Omega(\Delta^+ - a\Delta^-)^2]} \tag{20}$$

where $\Delta^+ = \omega + \delta$, $\Delta^- = -\omega + \delta$, $\gamma = (a+1)\Omega + (\Gamma/2)$
and $a = \frac{\Gamma}{\Omega} + k$.

In Fig. 7 we show the final SEDM line shapes including
analyzer thickness for different values of the various
parameters.

The superiority of the SEDM technique over the usual
approaches to the study of time-dependent effects can be
demonstrated by a direct comparison of the calculated results for the two types of experiments. We consider a
hypothetical case where the two peaks are 35 natural line
widths apart and calculate the line shapes for several
relaxation rates. Fig. 8 shows the calculated transmission spectra and the corresponding results appropriate
for the SEDM experiment. In Fig. 8 the curves show a
slight broadening which under experimental conditions may
be difficult to identify positively as due to relaxation,
since other effects such as sample thickness, collimation,
and field inhomogeneities cause a similar modification in
the line shape if not accounted for properly. However the

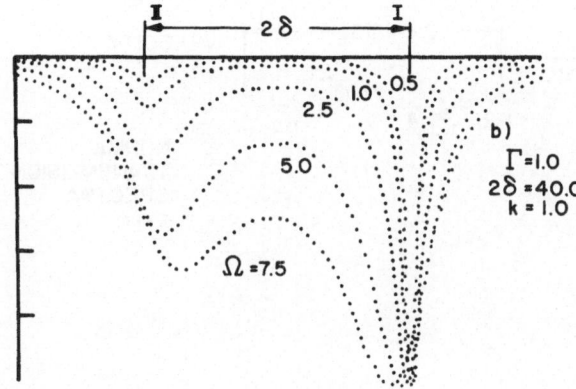

Fig. 7. Calculated SEDM relaxation spectra for different relaxation rates. The excitation occurs at line I and $2\delta/\Gamma = 40$.

SEDM results show the appearance of a second peak in all four cases and thus unambiguously determine the existence of relaxation in the scatterer.

We have used our SEDM techniques to study the Morin transition in $\alpha\text{-Fe}_2\text{O}_3$. The rhombohedral crystal hematite ($\alpha\text{-Fe}_2\text{O}_3$) is basically antiferromagnetic up to the Néel temperature ($T_N = 946°K$) and has a spin flip transition at $T_M = 263°K$ which is called the Morin transition[19,20]. Below T_M the antiferromagnetic axis is along the [111] direction while above T_M the spins lie in the [1̄11] basal plane. The electric field gradient (EFG) principal axis is along the [111] direction for temperatures both above and below T_M. This transition region has been studied experimentally[20-23] and also theoretically[24]. Ordinary Mössbauer experiments[25] show that above T_M the characteristic six line hyperfine pattern is different from the one observed at low temperatures (see Fig. 9). This is due to

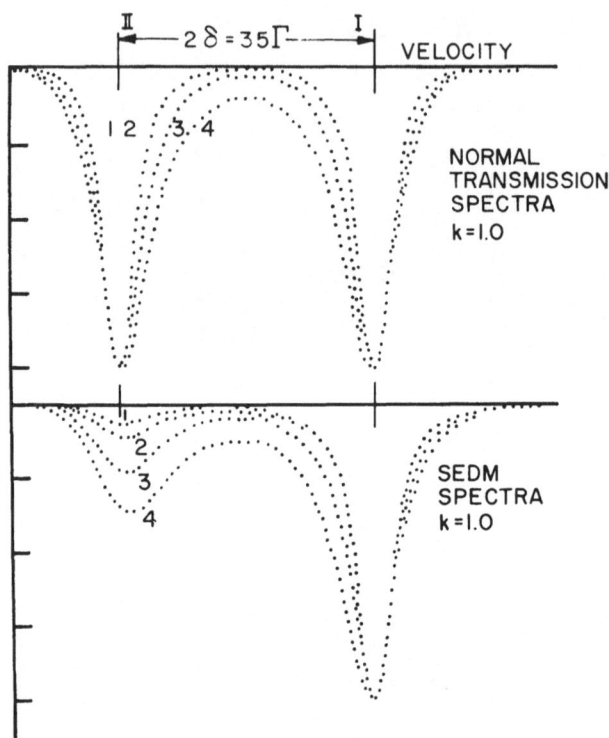

Fig. 8. Calculated relaxation spectra using (a) ordinary
Mössbauer transmission geometry, and (b) SEDM assuming
relaxation rates in energy units; (1) 0.25Γ, (2) 0.5Γ, (3)
1.25Γ, and (4) 2.5Γ.

the change in relative orientation of the EFG principal axis
and the direction and magnitude of the effective internal
magnetic field at the ^{57}Fe nuclei[25]. The difference in the
Mössbauer spectra can be characterized by noting the positions of the most energetic transition at room (8.34 mm/sec)
and liquid nitrogen (9.20 mm/sec) temperatures relative to
a Co - Cu source (see Fig. 9). Regular Mössbauer transmission experiments near T_M show the presence of both hyperfine patterns broadened and partially resolved. Is this

because both hyperfine interactions are present and the
observed spectrum is simply a superposition of the two, or
is each iron ion's spin flipping between the two possibilities? One could extract possible time dependent information
by applying standard relaxation theory to these ordinary
Mössbauer spectra[26]. Such an approach is very difficult in
this case because the two hyperfine patterns differ only
slightly from each other. SEDM experiments obviate this
difficulty.

Our SEDM experiments were performed using a scatterer
of 90% enriched α-Fe_2O_3 powder which had an effective
single line thickness parameter β = 350. Our first SEDM
experiments were performed with the α-Fe_2O_3 scatterer at
263.5°K. The CVD was set on the energy of the sixth line
at room temperature and the analyzer scanned over a
velocity range in the neighborhood of the sixth peak. The
results are given by the dots in Fig. 10. If these results
represent simply a superposition of two static hyperfine
patterns, they could be explained by using the results of
our time independent SEDM calculations. One would simply
have three contributions; the Rayleigh scattering, the
resonant contribution from exciting a peak at or near its
resonance (8.34 mm/sec), and another resonant contribution
from exciting a second peak off resonance. The important
point to remember is that the resonant contribution for
thick scatterers appears at the excitation energy. The
calculated result according to our time independent calculations is shown in Fig. 10 by the dashed line. The dashed
line can also be considered roughly as the sum of experimental spectra obtained by exciting the peak at room

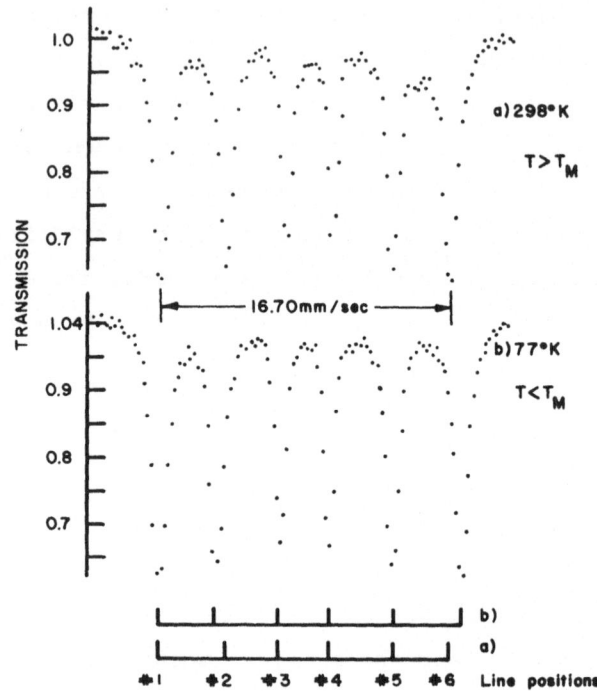

Fig. 9. Transmission spectra of α-Fe$_2$O$_3$ showing the two different hyperfine patterns (a) above, and (b) below T$_M$.

temperature near resonance and the peak at liquid nitrogen temperature off resonance. (The liquid nitrogen peak was actually slightly shifted in our calculations because of the temperature dependence of the internal magnetic field.) There is no way that the experimental data can be explained without invoking time dependent effects[4].

In order to apply our relaxation calculations, we needed to know the equilibrium populations of the two ionic states (Eq. 19). For this purpose a series of ordinary transmission runs were done using a thin (β=1) sample of

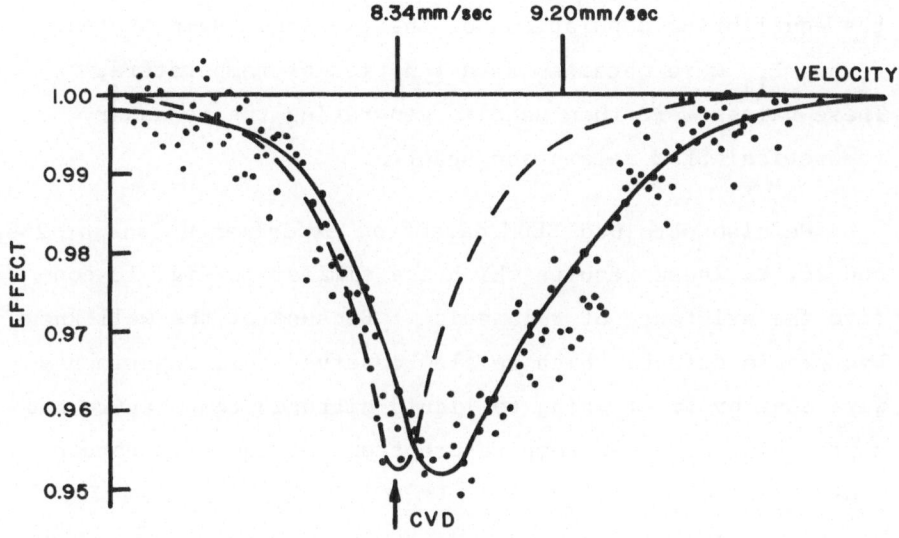

Fig. 10. Experimental SEDM spectrum of an enriched α-Fe$_2$O$_3$ scatterer at T = 263.5°K. The dashed line is the time independent result. The solid curve is the computed result for a relaxation time of 1.1 x 10^{-7} sec.

powdered α-Fe$_2$O$_3$. This thin sample was used in contrast with the thick one for SEDM, in order to obtain narrower lines and hence better energy resolution. The experimental data acquisition was set to run only over the velocity range of the sixth peak. The results show that at room temperature one observes a single line which corresponds to the room temperature hyperfine pattern. At lower temperatures a second line corresponding to a different hyperfine interaction appears and becomes increasingly more intense until at T$_M$ = 263.5°K, the two lines are of about equal intensity. A computer program was used to obtain the best two line Lorentzian fit to the experimental data. From these data

the equilibrium populations of the two ionic spin states
P_I and P_{II} were obtained as a function of temperature.
These values were then used in generating the necessary
theoretical SEDM relaxation spectra.

We also obtained SEDM data from experiments done at 259.5
and 261°K. These results which are similar to Fig. 10 confirm the existence of relaxation. Because of the well known
hysteresis effect, which we also observed, all experiments
were done by first going to liquid nitrogen temperature and
approaching each new temperature from the low temperature
side.

CONCLUSIONS

We have developed selective excitation double Mössbauer
(SEDM) calculations that apply to a "split" scatterer, and
include Rayleigh, incoherent electronic, and thickness
effects. These results can be used to fit experimental SEDM
data in the absence of relaxation processes. The important
point in these calculations is that the scattered radiation
has a peak <u>at</u> the excitation energy. We have verified these
calculations by doing SEDM experimental runs on enriched
iron powder scatterers at room temperature, and on an enriched $\alpha\text{-}Fe_2O_3$ powder scatterer at liquid nitrogen
temperature.

We have further demonstrated the advantage of the SEDM
technique in the study of relaxation processes, as compared
to the regular Mössbauer transmission procedure (see Fig.8).
We have applied these relaxation procedures to study the
Morin transition in $\alpha\text{-}Fe_2O_3$. Our results show that each
iron nucleus in $\alpha\text{-}Fe_2O_3$ near the Morin temperature is in a

fluctuating hyperfine field. The values of the relaxation
times at 263.5, 261, 259.5, and 258°K are respectively
1.1±0.2 x 10^{-7}, 2.3±0.4 x 10^{-7}, 2.9±0.6 x 10^{-7}, and
4.3±0.9 x 10^{-7} sec.

We have also observed relaxation effects in $\alpha-Fe_2O_3$
at room temperature using SEDM techniques although these results are not presented here. SEDM spectra of the same
material at liquid nitrogen temperature were consistent with
our time independent SEDM calculations. Transmission spectra
of $\alpha-Fe_2O_3$ at room temperature showed that the simple six-
line pattern, when looked at with improved resolution, is
composed of several patterns having slightly different
energy splittings. These spectra are all rather complex
and further work needs to be done before any definitive
statements can be made.

In general these SEDM procedures can be applied to
several problems of current interest. For example, an
electron hopping model has been used to explain the complex
Mössbauer patterns of magnetite as a function of temperature[27]. The Mössbauer spectrum of magnetite consists of
two superimposed six line hyperfine spectra corresponding
to A and B sites. The lines of the B site spectrum appear
to be broadened relative to those of the A site spectrum.
It is suggested that the broadening is due to an electron
exchange between the Fe^{3+} and Fe^{2+} ions at the B site.
This suggestion can be checked by using SEDM procedures.
Selective excitation of the least energetic transition in
the A site spectrum would result in a single SEDM peak
which could be fit using SEDM calculations without relaxation. However, excitation of a transition in the B site

spectrum would result in a line shape that would require the
incorporation of time dependent effects. If this electron
hopping model is incorrect i.e. if there are no local time
dependencies, then static SEDM theory would fit the result.

Also, the regular Mössbauer effect has been used to
study relaxation effects in such systems as Goethit
(α-Fe OOH)[28] and Ferrichrome A[29]. The analysis of such
experimental results has been characterized by Clauser and
Blume[30]. "In stochastic treatments of Mössbauer line shape,
the hyperfine interaction between the nuclear spin and the
electronic spin is replaced by an interaction between the
nuclear spin and randomly varying external magnetic and
electric fields. The randomly varying fields represent
the effect of a relaxing electronic spin on the nucleus.
This type of treatment is <u>only</u> valid, however, if the
electronic part of the hyperfine interaction field has no
off diagonal matrix elements, since an external magnetic
field has no such elements." However, using the SEDM procedures it is possible to tell if the interaction should be
represented by $A\vec{I} \cdot \vec{S}$ which would involve off diagonal
matrix elements. Thus SEDM results can check the generally
used effective field approximation.

REFERENCES

1. A.N. Artemev, G.V. Smirov, E.P. Stepanov, Sov. Phys.
   JETP <u>27</u>, 547 (1968).

2. W. Meisel, Monatsber, Deutschen Akd. Wiss. Berlin
   Ger. <u>11</u>, 355 (1969).

3. N.D. Heiman, J.C. Walker, and L. Pfeiffer, Phys. Rev.
   <u>184</u>, 281 (1969), and "Mössbauer Effect Methodology",
   Vol. 6, I.J. Gruverman, ed. (Plenum Press, New York,
   1971).

4. B. Balko and G.R. Hoy, Phys. Letters, (1974).

5. W. Heitler, Quantum Theory of Radiation (Clarendon Press, Oxford, 1954).

6. A.J.F. Boyle and H.E. Hall, Repts. Progr. in Phys. 25, 441 (1962).

7. P. Debrunner and R.J. Morrison, Rev. Sci. Instr. 36, 145 (1965).

8. B. Balko and G.R. Hoy, submitted for publication.

9. P.B. Moon, Proc. Phys. Soc. A63, 80 (1950).

10. P.J. Black, D.E. Evans, and D.A. O'Connor, Proc. Roy. Soc. (London) A 270, 168 (1962).

11. P.J. Black, G.Longworth, and D.A. O'Connor, Proc.Phys. Soc. 83, 937 (1964).

12. S.L. Ruby and J.M. Hicks, Rev. Sci. Instr. 33,27 (1962).

13. Yu Kagan, A.M. Afanasév, and I.P. Perstnev, Sov. Phys. JETP 27, 819 (1968) [Zh.Eksp.Theor.Fiz.54,1530 (1968)].

14. P.B. Moon, Proc.Roy. Soc. A263, 309 (1961).

15. M. Blume, in Hyperfine Structure and Nuclear Radiations, E. Matthias and D.A. Shirley Eds. (North Holland Publishers, Amsterdam, 1968) page 911.

16. P.W. Anderson, J.Phys. Soc. Japan 9, 316 (1954).

17. A. Abragam, Nuclear Magnetic Resonance (Oxford Univ. Press, London, 1961).

18. A. Paupolis, Probability, Random Variables and Stochastic Processes, (McGraw-Hill Co., 1965).

19. P.J. Flanders and J.P. Remeika, Phil. Mag. 11, 1271 (1965).

20. P. Imbert and A. Gerard, Compt. Rend. 257, 1054 (1963).

21.   S.T. Lin, Phys. Rev. 116, 1447 (1959).

22.   R.C. Liebermann and S.K. Banerjee, J.Appl. Phys. 41, 1414 (1970).

23.   S. Foner and S.J. Williamson, J. Appl. Phys. 36, 1154 (1965).

24.   J.O. Artman, J.C. Murphy, and S. Foner, Phys. Rev. 138, A912 (1965).

25.   F. Van der Woude, Phys. Stat. Solidi 17, 417 (1966).

26.   H.H. Wickmann, Mössbauer Effect Methodology, Vol. 2, I.J. Gruverman, ed. (Plenum Press, N.Y. 1968), p. 316.

27.   W. Kündig and R.S. Hargrove, Solid State Commun. 7, 223 (1969).

28.   F. Van der Woude and A.J. Dekker, Phys. Stat. Sol. 13, 181 (1966).

29.   H.H. Wickmann, M.P. Klein, and D.A. Shirley, Phys. Rev. 152, 345 (1966).

30.   M.J. Clauser and M. Blume, Phys. Rev. B3, 583 (1971).

LIST OF CONTRIBUTORS

D. G. Agresti, Department of Physics, University of Alabama in Birmingham, Birmingham, Alabama 35294

B. Balko, Physics Department, Boston University, Boston, Massachusetts 02215

U. Bäverstam, Institute of Physics, University of Stockholm, Stockholm, Sweden

M. Belton, Department of Physics, University of Alabama in Birmingham, Birmingham, Alabama 35294

C. Bohm, Institute of Physics, University of Stockholm, Stockholm, Sweden

P. Boolchand, University of Cincinnati, Cincinnati, Ohio 45221

J. B. Brown, Jr., Battelle Columbus Laboratories - Durham Operations, 3333 Chapel Hill Boulevard, Durham, North Carolina

J. P. deNeufville, Energy Conversion Devices, Inc. Troy, Michigan 48084

R. E. Detjen, Department of Physics and Liquid Crystal Institute, Kent State University, Kent, Ohio 44242

T. Ekdahl, Institute of Physics, University of Stockholm, Stockholm, Sweden

P. A. Flinn, Carnegie-Mellon University, Pittsburgh, Pennsylvania 15668

F. Y. Fradin, Argonne National Laboratory, Argonne, Illinois 60439

R. B. Frankel, Francis Bitter National Magnet
    Laboratory, Massachusetts Institute of Technology,
    Cambridge, Massachusetts 02139

J. M. Friedt, Laboratoire de Chimie Nucleaire, Centre
    de Recherches Nucleaires, 67200 Strasbourg,
    France

S. S. Hanna, Department of Physics, Stanford
    University, Stanford, California 94305

J. P. Hannon, Physics Department, Rice University,
    Houston, Texas 77001

G. R. Hoy, Physics Department, Boston University,
    Boston, Massachusetts 02215

C. W. Kimball, Northern Illinois University, DeKalb,
    Illinois 60115

E. König, Institutes of Physical Chemistry II and
    Physics II, University of Erlangen-Nürnberg,
    D-8520 Erlangen, Germany

D. Liljequist, Institute of Physics, University of
    Stockholm, Stockholm, Sweden

T. M. Lin, Department of Physics, Northern Illinois
    University, DeKalb, Illinois 60115

S. Long, Department of Physics, University of Alabama,
    in Birmingham, Birmingham, Alabama 35294

C. T. Luiskutty, Physics Department, University of
    Louisville, Louisville, Kentucky 40208

H. Maletta, Institut für Festkorperforschung der
    Kernforschungsanlage, 517 Jülich 1, Federal
    Republic of Germany

S. Mørup, Laboratory of Applied Physics II, Technical
    University of Denmark, DK 2800 Lyngby, Denmark

P. J. Ouseph, Physics Department, University of Louis-
    ville, Louisville, Kentucky 40208

R. S. Preston, Department of Physics, Northern
Illinois University, DeKalb, Illinois 60115

B. Ringstrom, Institute of Physics, University of
Stockholm, Stockholm, Sweden

G. Ritter, Institutes of Physical Chemistry II and
Physics II, University of Erlangen-Nürnberg,
D-8520 Erlangen, Germany

S. L. Ruby, Argonne National Laboratory, Argonne,
Illinois 60439

G. K. Shenoy, Laboratoire de Chimie Nucleaire, Centre
de Recherches Nucleaires, 67200 Strasbourg,
France

J. G. Stevens, University of North Carolina at
Asheville, Asheville, North Carolina 28804

V. E. Stevens, University of North Carolina at
Asheville, Asheville, North Carolina 28804

S. P. Taneja, Northern Illinois University, DeKalb,
Illinois 60115

G. T. Trammell, Physics Department, Rice University,
Houston, Texas 77001

B. B. Triplett, Department of Physics, Stanford
University, Stanford, California 94305

D. L. Uhrich, Department of Physics and Liquid Crystal
Institute, Kent State University, Kent, Ohio
44242

J. Webb, Division of Immunology and Rheumatology,
Department of Medicine, University of Alabama
in Birmingham, Birmingham, Alabama 35294

L. Weber, Northern Illinois University, DeKalb,
Illinois 60115

INDEX